Biodiversity Conservation in the Himalayas

Biodiversity Conservation in the Himalayas

B.L. Kaul
Former Principal
Government S.P.M.R. College of Commerce and Management,
University of Jammu
Jammu

2009
DAYA PUBLISHING HOUSE
Delhi - 110 035

ISBN 81-7035-612-1
ISBN 978-81-7035-612-7

Published by : **Daya Publishing House**
1123/74, Deva Ram Park
Tri Nagar, Delhi - 110 035
Phone: 27383999
Fax: (011) 23260116
e-mail : dayabooks@vsnl.com
website : www.dayabooks.com

Showroom : 4760-61/23, Ansari Road, Darya Ganj,
New Delhi - 110 002
Phone: 23245578, 23244987

Laser Typesetting : **Classic Computer Services**
Delhi - 110 035

Printed at : **Chawla Offset Printers**
Delhi - 110 052

PRINTED IN INDIA

To My Dearest

Anil, Kuldeep, Ahan & Upmanyu

To My Dearest

Aunt Kuldeep Ahun & Hamaayu

Foreword

It is a great pleasure writing this foreword to Professor B.L. Kaul's superb study on *Biodiversity Conservation in the Himalayas*. We all know that the Himalayan mountain system is home to a huge variety of flora and Fauna– but what we seem to have still not comprehended fully is the fact of its fragility. This whole eco-system is under threat from a constant onslaught of human greed.

Whether it is the foot of the hunter or a mountaineer or a gatherer of minor forest products, what is trampled, exploited and destroyed is the health of the Himalayas. If we do not wake up to the danger this irresponsible harvesting of the wealth of the system poses to the environment, the survival of the human race itself will be at stake. The delicate balance that nature maintains while creating its rich diversity, once disturbed irreparably, will have such disastrous consequences that a lay person cannot even imagine. Unfortunately, our policy-planners and executors are not really sensitive to the great apocalyptic future that awaits us, unless we act now.

It is against this background that Professor Kaul's book assumes great significance. Prof Kaul is, of course, well known as a great teacher and outstanding researcher. But in this book he has surpassed himself. It is essential reading for students, researchers, activists and, in fact, everyone seeking to save the Himalayas. I congratulate Professor Kaul on this splendid effort.

Amitabh Mattoo
Member, National Knowledge Commission
Professor of International Studies, Jawaharlal Nehru University, New Delhi

Preface

The colossal and awe-inspiring Himalayas have always been eulogized by the people of India as nature's marvel and the abode of gods.These mountains are truly one of the most outstanding yet fragile ecosystems of our planet. With lofty snow capped peaks, fast-flowing and awe-inspiring rivers, bubbling brooks, fascinating ethnic diversity, the mighty Himalayas are a thing of beauty. Many parts of the Himalayas are veritable mines of valuable biodiversity, thus making it the base for future food security as well as for livlihood security for the inhabitants of South-Asia.

The Himalayas which have served as an impregnable wall against the invasions from the north in the past are facing danger posed by none other than Indians themselves. These mountain chains are one of the most threatened life support systems on the earth today. The factors responsible for the degradation of the Himalayas are many. The most visible factor is the continuing deforestation and the resultant high intensity land-slides and floods. Unplanned developmental activities especially in mining, road construction and dam building using dynamite and explosives have added fuel to the fire. As if human vandalism was not enough nature also has had its share by way of earthquakes, neo-tectonic activities, cloud bursts, avalanches etc. to accentuate the ecological deterioration of the Himalayas with serious repercussions.

The desperate dependence of villagers living on the fragile mountains for agriculture, fuel wood and fodder for live-stock, has taken its own toll and resulted in denudation and virtual desecration of nature's charm and beauty. The beautiful high altitude lakes are shrinking due to pollution and eutrophication. Thus, denuded the Himalayas have been rendered ecologically useless and counter productive. As a corollary many among 8000 species of flowering plants (of which 40 per cent are endemic) and hundreds of species of wildlife in the Himalayas, are facing threat of extinction. The loss of biodiversity enhances our vulnerability to biotic and abiotic stresses thereby affecting the future of our natural food security system which depends to a great extent upon the ecological security of the Himalayas.

A more recent dimension to the threat to the Himalayas is the global warming. The warming of the Himalayas has increased noticeably over the last 50 years. As a consequence it is believed that the Himalayan glaciers are melting. The opinions on the threat to Indo-gangetic plain as a result of ice-melts in the Himalayas differ. According to one view the threat of glacial-melt is real and will result in flooding of glacial lakes, floods, land erosions and droughts. The other view is that the pace of glacial-melt is not so alarming and it is a myth that any drastic change will take place in near future.

The recent newspaper reports have added one more alarming dimension to the security of the Himalayas. It is reported that China has been dumping nuclear wastes in Tibet – the roof of the world. If these reports are correct the major rivers of South-Asia – Indus, Satluj, Ganges, Yumuna and Brahmputra will soon be carrying radioactive material thereby threatening the lives of millions of inhabitants of Pakistan, India, Nepal, Bhutan and Bangladesh.

Another serious threat comes from the opening up of the Himalayan peaks to trekkers and expeditions. These expeditions are causing environmental pollution in the Himalayas. Tourism has already emerged as a threat and is now synonymous with pollution and destruction of biodiversity in the Himalayas.

An unprecedented global demand for the herbal medicines, beauty creams and ointments has resulted in a fast deprivation of medicinal plants in the Himalayas. The threat is as much from across the Himalayas as on this side. This needs immediate check.

A more recent threat on a large scale that has emerged in the Himalayas is that of insurgency. The insurgents in North-West Pakistan, Nepal and Kashmir and North-Eastern states of India have played havoc not only with the human lives but also have been instrumental in wanton destruction of forests and wildlife. The hunting of the "Hangul" – the endemic Red Stag of Kashmir by gun yielding militants has resulted in drastic reduction of their numbers thereby seriously threatening their very existence.

In the face of threat to the existence of the Himalayas as they are coming mainly from the undue interference by man, ways and means have to be found to restore the pristine glory of the Himalayas. While all development is not bad in itself it has to go hand in hand with protection and conservation. Unfortunately, however, no major engineering project includes restoration cost in its financial projection. The need is to simultaneously repair all damage done to the land and water along with the progress of works.

The fret and fury of modernity has made the situation precarious and the need of the hour is to indulge in serious introspection and to evolve and suggest potent remedial measures to arrest the deterioration as quickly as possible. A clean, healthy and conducive Himalayan environment will radically better the health and living conditions of humans living in South-Asia.

How can we then safeguard the rich cultural, spiritual and genetic heritage that the Himalaya has blessed us with? The need of the hour is to promote balanced socio-economic development of the hill people even while ensuring preservation of its fragile ecological balance. This can be achieved only by a judicious blend of education, political will, professional skill balanced development and people's participation.

I am thankful to the contributors for sending across their research papers and hope that the volume will be well received. Thanks are also due to M/S Vinod Publishers and Distributors Jammu for allowing me to use some material from my earlier edited work titled, "Ecodegradation of Himalayas – Causes and Remedies", in the present volume.

Finally I must thank my wife Promila for her continued support and encouragement. I am also thankful to Mr. Rajesh Kumar for typing the manuscript with dedication and devotion. Last but not least I am thankful to Shri Anil Mittal of Daya Publishing House, New Delhi for his help and co-operation in bringing out this volume in the shortest possible time.

B.L. Kaul

Jammu, India

Contents

List of Contributors

Basu S.R.
Professor, Department of Geography, University of Kolkata, Kolkata, West Bengal

Bhan Daisy
Training Associate, Farmer Training Centre, Department of Agriculture, Jammu, J&K

Chauhan N.S.
Department of Forest Products, Dr. Y.S. Parmar University, Nauni, Solan (H.P).

Das Debashis
Professor Department of Geography, Visva-Bharati, Santiniketan, Birbhum, West Bengal–731235

Ganguly Raj
Independent Consultant, B25, OCS Apartment, Mayur Vihar, Phase–I, New Delhi–110 091; Consultant, FAO, New Delhi

Gupta R.D.
Ex-Chief Scientist-Cum Associate Dean, Sher-e-Kashmir University of Agriculture Sciences and Tecnology, R.S. Pura, Jammu.

Kaul B.L.
Former Principal, S.P.M.R. College of Commerce and Management, Jammu
Present Address: 186, Upper Laxmi Nagar, Sarwal, Jammu–180 005
E-mail: blkaul@gmail.com, kaul.bansilal@yahoo.com

Kaul M.N.
Professor, Department of Geography, University of Jammu, Jammu, J&K

Kher Deepak
Division of Soil Science and Agricultural Chemistry, SKUAST-J, Main Campus Chatha, Jammu–180 009

Macoo S.N.
Former Divisional Soil Conservation Officer, Soil Conservation Department (J&K)
Present Address: 1033, Apna Villa Aptts, Plot 23, Sector 10, Dwarka, New Delhi–110 075

Lamo N.D.
Senior Lecturer, Department of Geography, Siliguri College, Siliguri (W.B.)

Madhumita Sinha
Department of Geography, University of Kolkata, Kolkata, West Bengal

Nabanita Sen Gupta
Department of Geography, University of Kolkata, Kolkata, West Bengal

Nandini Dhara
Department of Geography, University of Kolkata, Kolkata, West Bengal

Parimoo A.K.
Chief Engineer, Design Directorate, J&K, Jammu
Present Address: Priya Darshani Lane, Bohri, Jammu

Roy Partha Pratim
Department of Geography, Siliguri College, Siliguri, West Bengal

Sharma O.P., *I.F.S.*
Joint Director, State Forest Research Institute, Jammu

Sen Jyotirmoy
Reader and Head, Department of Geography, Siliguri College, Siliguri, West Bengal

Teli B.L.
Department of Geography, H.N.B. University, Pauri, Garhwal, Uttarakhand

Thakur Sanjeev
Department of Tree Improvement, Dr. Y.S. Parmar University, Nauni, Solan (H.P)

Tiku J.L.
Former Managing Director, Milk Federation, Jammu

Verma Anil Kumar
Selection Grade Lecturer, Department of Zoology, Govt. Post Graduate College for Women, Gandhi Nagar, Jammu

Chapter 1

Climate Change and its Impact on Glaciers of NW Himalaya: An Overview

M.N. Kaul

ABSTRACT

The glaciological studies carried out in NW Himalaya during last twenty years indicate that the glaciers are receding and thinning at a very slow rate and as such are not showing any alarming change as reported in alpine glaciers. Further while comparing meteorological data collected by automatic weather station reveal that glaciers do not show any immediate response to climatic parameters particularly climatic data collected at a glacier site during 1994 to 2003. During 2001-2002 the glacier (Naradu) received snowfall one and half times that of 2000-2001 and 2002-2003, similarly the maximum mean solar radiation was 470 watt/m^2 2002 and 510 watt/m^2 2003 and maximum temperature recorded was 22° C, 21.5° C and 19.7° C respectively during the above mentioned period. The results clearly reveal that though during 2001-2002 glacier received highest snowfall but still the retreat was normal (-4.9m) and also the retreat does not show any relation with the temperature as well as solar radiation. The glaciers showed a retreat of 4 to 8.4m in SE of NW Himalaya and 2.5 to 5.7m in NW Himalaya. Further the result of 13 glaciers of NW Himalaya has been in phases during last century. Most of the glaciers have retreat at the rate of 10 to 25m up to 1990, since then they have been showing gradual reduction in retreat. Even the two largest glaciers Siachen (Nubra valley) and Durung Drung (Zanskar valley) have been showing moderate to low retreat since 1996. This clearly indicates that glaciers in NW Himalaya are not responding to so to say global warming effect. Even the Global Climate General Circulatory Model (GCGCM) predict little warming or alternate cycle of warming and cool climate after thirty years. According to them global cooling would take place from 2006 to 2040 and warming from 2040 to 2075 like wise.

Introduction

The atmosphere consists of nitrogen (78.084 per cent), oxygen (20.964 per cent), Argon (0934 per cent), carbon dioxide (0.0314 per cent) and other trace gases including methane, water vapor, dust, pollen grains, pollutants like oxides of carbon, nitrogen and sulphur, hydrogen fluoride, lead fumes hydro-carbons and secondary products. Many of the substances are added by anthropogenic sources and are harmful to life and therefore are grouped as pollutants. Most of the pollutants are produced in lower altitudes by emission of the gases from the industries, automobiles and by other man induced actions. These pollutants have degraded the atmosphere and causing warming effects in the lower atmosphere. During last fifty years a lot of sensation has been created by media about the depletion of atmosphere by gradual increase of green house gases like carbon dioxide, methane, chloro-fluro carbon and nitrous oxide resulting in "Global Warming". Even the buzzword of today is global warming. This has very serious implication towards global climatic change leading to far reaching consequences of snow melt, rising sea level, drowning of coastal low lands, altered rainfall and evaporation, creating of new plant diseases and other problems. As per the ground truth such realities are not occurring on a alarming rate rather they are dished, day in and day out through media created by certain alarmist elements.

Global Warming and Climate Change

The global warming theory was propounded in nineteen eighties by an eminent NASA scientist James Hansen, since then it has generated lot of debate in the scientific communities. A number of global climate general circulation models have been evolved regarding sensitivity of change in global temperature. Three major schools of thoughts have emerged, one school predicting lower rate of climate change as temperature is likely to increase (1° to 3° C) due to doubling of carbon dioxide concentration in atmosphere. While the other group acknowledges a 10° C increase. The third group proposes an alternate cycle of warm and cold temperature after every 30 years of period due to sun spot and solar irradiance rather than the carbon dioxide concentration in atmosphere. Even propounder of the global warming admitted in the year 2005 that it was impossible to come out with any reliable climate model because scientists do understand that once a temperature shoots how does other elements behave that are constant at present in the atmosphere.

The climatologists fully understand that if concentration of carbon dioxide increases in atmosphere it is going to increase surface air temperature provided "all else" is held constant. The problem becomes lot more complicated in the real world when we consider that "all else" cannot be held constant and there are lot more changes occurring at any one time than the concentration of carbon dioxide. Once the temperature of earth starts increasing immediately afterwards the moisture level as well as cloud pattern of the lower atmosphere changes due to addition of moisture induced by warming effect. Consequently low clouds would retard or even reverse the warming leading to negative feed back. If these feed backs are simulated in global climate general circulation models (GCMS). Some GCMS predict little warming for doubling carbon dioxide and others predict substantial warming for same change in atmospheric conditions. The changes in carbon dioxide in the real world would almost certainly be associated with other changes in atmosphere like sulphur dioxide, mineral aerosol dust, ozone, black carbon, who knows what else would vary through time and complicate all "else held constant" picture.

The United Nations Inter Government Panel on Climate Change (IPCC) suggest that the sensitivity is between 0.48 and 1.4° K per one watt per square meter equals to global warming of 2.0° C to 4.5° C for doubling of carbon dioxide concentration.

A team of New-Mexico's Loss Alamos National Laboratory and Switzerland's Institute for atmospheric and climate science re-examined the extracted Ice Core of Vostok Antarctica, for assessing the temperature carbon dioxide, methane and dust records for nearly half a million years (Chylek and Lohman, 2008). The core studies reveal that around 42,000 years back earth experience cooling that lasted up to 15,000 years ago when earth relatively warmed. The scientists assess the sensitivity parameters of atmosphere from 15,000 years to the present time by evaluating the concentration of the carbon, methane and dust particles from the core samples. They came out with the finding that the sensitivity range of the atmosphere is far-far lesser (0.5 to 2) than the IPCC sensitivity range. There findings have been further corroborated by other group of scientists indicating far lesser increase in temperature due to doubling of carbon dioxide (Miura *et al.*, 2005; Wyant *et al.*, 2006; Chylek *et al.*, 2007 and Schwartz 2007). As per the recent findings there is a less temperature rise, and the less chance of major disruption.

The third group of scientists came out with the finding that the carbon dioxide cannot be the cause of climate warming. According to them prior to 1945 the atmospheric carbon did not increase significantly but still warming period was observed. After 1945 there was a rise of 80 per cent carbon dioxide but the temperature for next 30 years cooled 0.5°C in the northern hemi-sphere and 0.2°C in southern hemisphere instead of global warming. Glaciers have recorded a cyclic pattern of 25 to 35 years of warm and cool cycle for the past 400 years only the last one could possibly be attributed to the atmospheric carbon dioxide. During Little Ice Age temperature dropped to 4°C lasting few centuries up to 1,000 AD. Thereafter global temperature rose several degrees up to medieval period then dropped again. According to them the global warming curve would suggest cooling from 2006 to 2040, warming from 2040 to 2075 and cooling from 2075 to 2100 with total warming of only 0.3° C.

Himalayan Glacier Geometry

Himalaya, the youngest mountain range, have a dramatic landscape with snow, glaciers and surging river system for about 2400 kms spreading between Indus in the NW and Brahmaputra in the NE. The Himalaya, a result of geological events that goes back to almost 60 m yr, attained its present elevation since 15 m yr that led to the initiation and establishment of monsoon system in Asia. Pleistocene has been the time period when pulsating uplifts resulted in pushing Himalaya further up that resulted in the precipitation to be in the form of snow around the year giving rise to large glaciers (Valdiya, 1993). Glaciers are dynamic bodies of ice that move in Himalaya under the influence of variable temperature gradient. The temperature of ice plays a major role in its movement and morphological activity. The glaciers in Himalaya are both of warm and cold type. The warm type of glacier is confined to narrow valleys of Himachal and Kashmir Himalaya, where temperatures of ice is close to melting point throughout its thickness except the upper accumulation part where temperature is far below freezing conditions and melt water is at the base of debris covered glaciers. The cold glaciers are found in cold desert of Ladakh. They are characterized by temperature below melting point throughout their thickness. The surface and basal melting occurs in short summer ablation season and supra glacial streams temporarily appear in the lower ablation zone. The glaciers are, by and large, clean and are confined to broad valleys.

Himalayan glaciers comprise nearly 30 per cent of all the glaciers outside Polar Region and cover an area 20 times of the glaciers of Alps (Wissman, 1959). According to the Inventory on Himalayan glaciers prepared by Geological Survey of India, there are altogether 5218 glaciers extending from Kashmir in the west and Sikkim in the east. 3200 glaciers are confined to Himachal, Kashmir and Ladakh Himalaya along the major watersheds of Sutlej (616 glaciers), Beas (277 glaciers), Ravi (172 glaciers), Chenab (1278 glaciers), Jehlum (133 glaciers) and Indus-Suru (697 glaciers). The NW Himalayan glaciers contribute nearly 70 per cent of annual flow of melt water and east Himalaya contributes relatively less due to high snow line and intense monsoon. On an average, a Himalaya glacier produces 0.1km^3 water during summer season. Snow and glacier melt contribute 400 to 800 km^3 of water along with 100 tons/ ha/year of sediment which is of great economic value (Bahadur 2000). Mountains are source for over 50 per cent of glob's river. The volume of ice in a glacier, and correspondingly its surface area, thickness and length is determined by the balance between inputs and outputs that are largely controlled by temperature, humidity, wind speed, slope angle and ice albedo (Fitzharris *et al.*, 1996).

The precipitously descending snow and ice covered high peaks and ridges to glacier basins and steep relief is a cause for numerous ice falls, heavy crevassing, and massive debris load. As a consequence, some glaciers surge by block schooled motion, whereas majority of glaciers show gradual fluctuation. Some glaciers, particularly avalanche type, behave erratic and fluctuate rapidly causing temporary damming of rivers and thereby threaten the safety of recently built highway and reservoirs, such as that of Sutlej basin.

Further, the high relief, steepness of slope and debris accumulation provide an overwhelming sense of instability of slopes, mass movement and occurrence of catastrophic events that could lead to widespread slope failures leading to temporary blockade of rivers and subsequent floods. Occurrence of such hazards in high mountains leads to catastrophes that exert a substantial and consistent influence on society in the downstream.

Glacier Evidence and Climate Change

The mountain glaciers, such as those that exist at higher elevations in mid latitudes and tropics are particularly sensitive indicators of climate change; many studies have been carried out in different parts of the world on their response to climate change due to global warming in the 20th century. They experience a general state of advance with a maximum reaching some times to later part of 19th century and steady retreat thereafter. On the global scale the studies have shown that glacier wastage has been pervasive during the last century with small glaciers disappearing in mid latitudes where as larger one shrinking slowly and in polar region warming considerable. In Himalaya because of ruggedness of terrain and its inaccessibility very few glaciers have been studied. A few studies carried out by our team during last two decades particularly Kohali, Shisram, Machoie (Kashmir)Durung Drung, Kangriz (Zanskar Ladakh), Kol (Jammu) and Naradu (Himachal Pradesh) Glaciers in NW Himalaya. In the light of the Global Warming concept the glaciers of the NW Himalaya have been studied in detail to understand the response of global warming in terms of variation in length, mass balance, impact of micro climatological variables on the glacier body.

The glaciers are important barometers to assess the recent climatic change as they are the repositories of past climate. The main source of data is imprinted in climatic flora zone which monitors the fluctuation of the glacier. As such glacier is an important special system in which two main sub system: glacier mass system and glacier process- convincingly demonstrated as are positive and negative feed back processes which sustain the continuity of mass balance. The two zones of ablation and accumulation having different sites is bound to add to the mass of glacier and vice versa. The advance and the retreat of glacier are based upon snow gain through accumulation and snow loss through ablation. The

Dynamics of the glacier are influenced by variety of local and regional factors like amount of precipitation and temperature regime. Changes in atmospheric temperature and changes in amount and type of precipitation are bound to affect glacier health, but as a result glacier does not necessarily advance during colder weather or retreat during warmer weather. The glaciers are self supporting system they receive snow fall during the summer season due to sublimation of glacier surface due to local environment that causes precipitation in higher reaches and it sustains the glacier. Local environment near the glacier front (Snout) may sometimes show closeness with climatic parameters but it may be mere coincidence. As such it will be a prudent to come to a definite conclusion that glaciers in Himalayas are subjected to stress due to climatic change induced by global warming in context of Alpine glaciers. We should not equate Himalayan glaciers with the Alpine one as the Himalayan glaciers extend upto the altitude of 6200m in comparison to alpine 3000m. As such there is a relative difference of 3200m between the two and secondlay Himalayan Glaciers have a steep slopes and their relative relief goes upto the altitude of 2500m. Due to high relative relief the northwestern Himalayan peaks become "warm island" and at the same time become the centre of condensation due to sublimated air mass from the glacier ice to become a "moist island". There are two zones of high precipitation having origin to two different air masses, one generated by local weather condition and second generated by western disturbances and monsoons. High precipitation at higher altitude of higher Himalaya is most favourable for maintenance of glacier. This precipitation usually takes place during the summer season particularly along the accumulation zone and help in cementing the glacier and reducing the albedo affect.

The topographic development in NW Himalaya is inherently polygenetic in nature resulting in the existence of various processes, such as slow process of glaciation and fast of fluvial. All processes play important role from time to time. The change in climate is rapid with respect to altitude as well as aspect, such as rain shadow effect (caused by high mountain ranges), leading to sub-humid climate in southeast (Himachal and Kashmir Valley) and cold arid climate in northwest (Ladakh) Himalaya. In Himalaya, rapid change in atmospheric conditions is because of steady decrease in thickness of atmosphere that leads to sharp increase in ultra violet radiation causing sunburns. The density of atmosphere decreases the concentration of water vapors and carbon dioxide that in turn heats and cools ground surface rapidly and leads to acute cold condition in NW Himalaya.

Results and Discussion

The NW Himalaya is intrinsically linked to atmospheric circulation caused due to western disturbances that results in maximum deterioration of weather causing heavy snowfall that in turn give rise to extreme cold conditions favorable for the growth and sustenance of glaciers. The western disturbance is caused through thermo-dynamic effect of Himalaya on the periodicity and incidence of depression. The depression, with a life of barely 5 days, usually takes 8 to 10 days from its source to reach Himalaya. During the winter season (December-January) temperature ranges between -10° C and -25° C (Naradu) in southeastern flanks while in northwestern flank (Zanskar and Nubra) it ranges between -20°C and -45°C. During the summer season (July to September) the mean temperature ranges between 10°C and 15°C with July being the hottest (25°C) and solar radiation peaks to 600 watts/m^2. During summer season monsoonal winds do penetrate up to an altitude of 3500 m all along the southeastern flanks of the NW Himalaya. It pores heavily in the form of cloud burst and point rainfall, that causes flash floods in mountain regions of NW Himalaya.

The extremes of most of meteorological variables like snowfall, wind, solar radiation, ambient temperature and rainfall can be a cause for hazard. In addition, compounding of variables, such as low temperature combines with high winds to create a wind chill, snow storms, and avalanche etc., greatly increases threat for catastrophe. The nature and intensity of hazards in NW Himalaya vary considerably through length and breadth of Himalaya. In the southeast, the intensity of monsoon winds trigger mass movement and cloud burst processes resulting in flash floods. The western disturbances invade the NW Himalaya with a chilly winds followed by heavy snowfall that causes activation of ice slab avalanches, snow blizzards, mass movement processes, etc. in the mountain region between an altitude of 4000m and 2500m.

Glacier Fluctuation

The process of advance and recession of glacier results in the occupation and evacuation of area by glaciers through time and space. The process helps create hazardous moraine dammed lakes behind unstable ice cored moraine having potential to burst and cause flash floods. The glaciers in NW Himalaya exhibit accumulation area ratio ranging between 55 per cent and 65 per cent and indicating that majority of the glaciers are receding and depleting (Sangewar 1995). The rate of retreat is quite variable depending up on the location and aspect of glacier basin. Present day snout of main Baspa group with respect to one in 1965 topographic sheet confirms the horizontal retreat of 320 m in 33 years and remnant moraine about 5m to 6 m higher than the present glacier snout suggest vertical retreat of ice (Sangewar *et al.*, 1996). The glaciers retreat has been at variable rate, from southeast to northwest of NW Himalaya, (Table 1.1). All these glaciers have been retreating in phases during the last century most of them have retreated at a moderate rate of 10m to 25m up to 1990's, since then they have been showing gradual reduction in retreat. The two largest glaciers Siachen (Nubra valley) and Durung Drung (Zanskar valley) have been showing moderate to low retreat since 1996. Some of the glaciers like Machoi and Durung Drung are showing very low retreat. Overall all the glaciers are showing consistent reduction in size and comparative lower rate of retreat during last seven years in contrast to global warming phenomena. Further the Table 1.1 shows that the smaller glaciers like Kolahoi and Machoi (Kashmir valley) and Naradu (Himachal Pradesh) are also behaving like large glaciers. Further details study was carried out in Naradu glacier in Himachal for the mass balance estimation. The study reveals that glacier has a net balance of -1.28 x 10^6 m^3 (2000-2001), -0.329 x 10^6 m^3 (2001-2002) and -1.182 x 10^6 m^3 (2002-2003) showing secular retreat of 4.2m. 4.9m to 5.4m respectively. Whereas the pattern of snow accumulation during 2000-2001, 2001-2002 and 2002-2003 indicate that in the year 2001-2002 the total snow accumulation was 97.23 inches out of which 71 inches was received in the form of snowfall alone. It was one and half times more than the balance year of 2000-2001(55.35") and 2002-2003 (52.99") and similarly maximum mean daily solar radiation was 470 watts/m^2 (2001), 436 watts/m^2 (2002) and 510 watts/m^2 (2003) and the maximum temperature recorded by automatic weather station was 22° C (2001) and 21.5° C (2002) and 19.7° C (2003). Whereas the lowest temperature recorded was somewhat stagnant during these three years. The above results clearly reveals that though during 2001-2002 the glacier received highest snowfall but still the retreat was normal (- 4.9m) and also showing no relation with maximum and minimum temperatures. Thereby indicating the glacier does not show any immediate response to climatic parameters and further corroborating that there is no impact of global warming on glacier retreat.

Table 1.1: Average Retreat in the NW Himalayan Glaciers

Glaciers	*Period*	*Years*	*Retreat*	*Average Retreat mts./Year*
Kolahoi	1857–1983	126	1792	14.1
Kolahoi	1983–1987	4	15.2	3.8
Machoi	1906–1957	51	457	9.0
Machoi	1982–1988	8	Advance	2.0
Machoi	2002–2005		No change	
Shishram (Liddar)	1923–1983	60	501	8.4
Shishram (Liddar)	1983–1987	4	15.9	4.0
Nehanar	1976–1977	1	2	2.0
Siachen Nubra	1889–1935	45	552	13
Siachen Nubra	1956–1971	15	300	20
Siachen Nubra	1977–1990	13	312	24
Siachen Nubra	1990–1996	7	196	28
Siachen Nubra	1996–2001	6	150	25
Siachen Nubra	2001–2004	4	60	15
Siachen Nubra	2004–2006	2	26	13
Durung Drung (Zanskar)	2004–2007	3	10.2	3.4
Naradu (HP)	1962–1978	16	Advance	8.25
Naradu (HP)	1978–1994	16	24	15
Naradu (HP)	2001–2003	4	19.2	4.8

Observation

The NW Himalayan glaciers are showing secular retreat consistent with a retreat that began at the end of last glacial cycle. The studies carried out on 175 glaciers in Miyar valley of Chenab basin (H P) by satellite data show that large number of glaciers have a negative balance whereas a very few small glaciers show positive balance and large number of niche glaciers does not show any accumulation, (Bahuguna *et al.*, 2008). The results also indicate that small niche glaciers retreat faster in response to climate variation. Studies carried on glaciers of Zanskar basin reveal that glaciers are retreating at very slow rate. There has been net reduction of 4.23 per cent in glacier area and 3 per cent reduction in peri-glacial area of the Zanskar landscape in last 35 years (1965-2001) (Koul *et al.*, 2007). The glaciers in the region are retreating at normal rate in view of climate fluctuation. This also corroborate with mean daily discharge of Sutlej, Chenab, Beas and Ravi whose headwaters contribute 60 per cent of their total flow from about 18 per cent of catchment area.

The Himalayan glaciers have been receding and thinning at a very slow rate and as such is not showing any alarming change as reported in alpine glaciers. In Alps, the glaciers have reduced by 26 per cent during last one century (Muller *et al.*, 1976). This has resulted in the perturbation of natural processes like moraine dam failures, excessive retreat of glaciers, and, decreasing the nature of permafrost in the alpine region (Evans S. G. *et al.*, 1994; E. Taylor and B. Taylor 1997). Similarly, in Tibetan Plateau the rise of temperature at the rate of 0.3°C per ten year (Zhang and Tang 2000) has been, the cause for the formation and bursting of Pareechu lake and also triggering the landslides in the Tibetan region. Such phenomenons are not yet observed in NW Himalaya during last 20 years. The studies, so far, in Himalaya on the response of glacier and glacier related hazard to climate change need to be strengthened by way of:

1. Long term monitoring and analysis of indicators of environmental change in Himalayan region with a particular focus on cryospheric indicators, watershed hydrology and terrestrial ecosystem; and
2. Integrated model-based studies of environmental change in different segments of Himalaya based on the interaction between land surfaces and atmosphere.

It is yet to understand that how various natural modes of climatic variability, such as *El Nino*/Southern Oscillation (ENSO) or the North Atlantic Oscillation (NAO) perturb the mean precipitation pattern on the time scales ranging from seasons to decades in Himalayan region.

References

Bahadur, J., 2000. Water resource management in the Himalayan region. In: *Proc. Int. Conf. on Managing Natural Resources for Sustainable Agriculture Production in the 21st Century*. New Delhi

Bahuguna, I.M., Rathre, P P. and Kulkarni, A.V., 2008. Application of multitemporal Resources Sat–1 data to assess glacier mass balance: A case study of Miyar basin. In: *Proc. Int. Workshop on Snow, Ice, Glacier and Avalanches*, CSRE, IIT Bombay, p. 350–352.

Beniston, M., 2003. Climate change in mountain regions: a review of possible impacts. *Climatic Change*, 59: 5–31.

Cheylek, P. and Lohman, 2008. Aerosol radiative forcing and climate sensitivity deduced from last glacier maximum to Holocene transition" *Geophysical Research Letters*, p. 35.

Evans, S.G. and Clague, J.J., 1994. Recent climatic change and catastrophic geomorphic processes in mountain environments. *Geomorphology*, 10(1–4): 107–128.

Feng, S., Yao,T.D. and Jiang, H., 2001. Temperature variation over Shinghai-Xizang in past 60 years. *Plateau Meteorl.*, 20: 105–108.

Fitzharris, B.B., Allison, I., Braithwaite, R.J., Brown, J., Foehn, P., Haeberli, W., Higuchi, K., Kotlyakov, V.M., Prowse, T.D., Rinaldi, C.A., Wadhams, P., Woo, M.K. and Youyo Xie, 1996. The cryosphere: Changes and their impacts. In: *Second Assessment Report of the IPCC*. Cambridge University Press, 5: 241–265.

Kaul, M.N., Ganjoo, R.K., Dhinwa, P.S., Pathan, S.K. and Ajai, 2007. Use of geo-informatics for preparation of a desertification ststus map of Stod valley (1F4C2), Padam (Zanskar), District Kargil, J&K State. *Indian Society of Geomatics Newsletter*, 13(2 and 3): 27–34.

Muller, F., Caflisch, T. and Muller, G., 1976. Firm und ESI der Schweizer Alpen Publs. No. 57, Geographical Institute ETH Zurich.

Pascoe, E.H., 1964. *A Manual of Geology of India and Burma*, 3: 2079–2086, Madras.

Raina, V.K., 2006. Glacier retreat and global warming: Review. In: *19th IGI Conference on Himalaya to India Ocean and Landscape Change*, RCFOHG, Jammu University.

Sangewar, C.V., 1995. Glacier inventory (Level 3 and 4) of Sutlej Catchment, Kinnaur district, Himachal Pradesh. *Geological Survey of India, Spl. Pub.*, 21(2): 321–325.

Sangewar, C.V., Mukherji, B.P. and Singh, R.K., 1996. Glacier regime of Shaune Garang in drought year (1987), Baspa Valley, Kinnaur district, Himachal Pradesh. *Geol. Surv. India Spl. Publ.*, 21(2): 281–284.

Shang, Y., Yangz, Li, Liu, L., Liao, D.Q. and Wang, Y., 2003. A Super large landslide in Tibet in 2000: Background occurrence, disaster and origin. *Geomorph.*, 54: 225–243.

Taylor, E. and Taylor, B., 1997. Impact of climatic change on catastrophic geomorphic processes in the mountains of British Columbia Vol. 1: Canada country study: Climate impact and adaptation, British Columbia Ministry of Environment lands and Parks, Vancouver Canada (Edited Vol.).

Valdiaya, K.S., 1993. *Dynamic Himalaya*. University Press (India) Pvt. Ltd.

Valdiya, K.S., 1993. Uplift and geomorphic rejuvenation of Himalaya in quaternary period. *Current Science*, 64(11): 873–885.

Wissman, H. Von, 1959. Die Heutige Vergletschcrung and Schneegrenze in Hoschasien Abh Wissenchaft Lit. *Math-Naline,* 14: 1101–1407.

Zhang, D.L. and Tang, X.P., 2000. Analysis of change in temperature and rainfall in Tibet past 40 years. *Tibet Sec. Technot.*, 2: 62–66.

Chapter 2

Ecodegradation of the Himalayan State of Jammu and Kashmir: A Review

J.L. Tiku

ABSTRACT

The ecodegradation of the Indian State of Jammu and Kashmir in North-Western Himalaya is a matter of great concern. It is fast approaching a stage where the life is threatened by shrinking wetlands, widespread pollution of air and water, disappearing forest cover and wildlife and receding glaciers due to global warming, thanks to human interference. Climate changes in the Himalayas are being increasingly witnessed which will lead to unimaginable consequences. Increase in the livestock is also emerging as one of the causes of the degradation of the hill and mountain ecosystems.

Introduction

"Yeli Wun Poshi, Teli Poshi Aun." These are the prophetic words of the venerated poet Nund Reshi, the ascetic and Alamdari Kashmir. An abundant and luxuriant forest is a precursor for an excellent cereral crop. A dense forest is source of a divine blessing to the mankind in facilitating a rich rainfall, preservation of moisture, feeding springs and glaciers, creating luxuriant herbage, pastures and shrubs. It also supports a healthy and varied wildlife. However, Table 2.3 giving district-wise forest area 2004–05 gives a gloomy

picture. Against an ideal forest cover of 33 per cent the state forest area has shrunk to 14.56 per cent. It was around 15.2 per cent in early nineties. The highest density of forests is in Doda district, followed by Baramulla, Udhampur and Anantnag. Leh and Kargil have almost negligible forest cover because of adverse climate. Literally, a forest is a large tract of land covered with trees mixed with pasture and undergrowth. In 1996 Forest Act, it is left to States to come out with a definition of Forests. Thus in Madhya Pradesh 200 trees per hectare can be termed as forest and in Ladakh steppes it can be a barren landscape with a scattered shrub cover as a forest. All state governments are to file a report about it to the Ministry of Environment. While, it may take some time to come out with a final report, the fact remains that forest cover is rapidly declining in the Country as a whole. A century earlier the forest cover for the country was 40 per cent. It got reduced to 20 per cent in 1951 and in 1997 it was 19 per cent.

Varied Physiographic Conditions and Climates

The Indian subcontinent presents the most varied physiographic and climatic conditions present anywhere in the World. Ecological diversity has resulted in one of the richest source of flora and fauna in our country. India has the cold desert vegetation in Ladakh in J&K state, Alpine meadows in Himalayas, with Deodar and pine forests dotting lower Himalayas and tropical rain forests of Assam and Meghalaya and Western Ghats. The Xerophytic vegetation of the Thar desert of Rajasthan desert lends a unique position to the country in having so diverse vegetation. The Sal forests of the Gangetic plains and the Sunderbans with a massive expanse of marshy trees swaying to and fro with tides lends a signiificant distinction to this area of Bengal with Royal Tiger inhabiting this area in considerable numbers. All these vegetational zone have interesting plant species. It is estimated that there are 15000 known floral species found in India. The North east region alone have 7000 species. Since the dawn of Independence 50 lakh hectares of forest land have been lost due to varied causes. The undergrowth shrubs, herbs and creepers which survive under tall deodars, pine or sal check the soil erosion and are the store houses of medicinal plants. The recent spurt in resorting to commercial forestry have resulted in reduction of hundreds of plant species in providing space to the sown man made forests. The undergrowth is ruthlessly weeded out to provide space to the sown forest tree which in recent past was the moisture depleting Eucalyptus. Such forests are in fact ecological deserts. These manmade forests harbour a very small percentage of indigenous flora and fauna. Most of the wild animals, birds and insects depend on the underbrush for food, shelter and protection. In the man made forests there are large acreage of eucalyptus and teak and very few denizens of forests because the under shrub is lacking. Rain forests of Assam and north east had 7000 species of plants which was half of the country's floral wealth. The massive urbanization, deforestation and industrialization pose a serious threat to them. Assam and Meghalaya are extremely rich in orchid species but wanton collection by hobbyists poses a threat to many rare species and no sanctuaries are being created to allow unhindered growth and protection of these rich floral orchids.

Cheshmashahi Hills are getting back their Verdant Cover

Cheshmashahi is famous in Srinagar for its Spring and beautiful garden. It has a unique observatory and seminary attributed to Dara Shiko a secular Mughal prince and son of King Shah Jehan. It also houses a retreat of famous spiritual saint poetess Rup Bhavani known as Cheshmah Sahibi. It also houses the residence of Governor of J&K state. There are a number of tourist huts around and milk supply scheme. Earlier it also had a state run central cattle breeding farm, which has now been shifted somewhere else. I had the privilege of working at the government run Farm in 1971 which had a large herd of cattle both exotic and other breeds. I again had an occasion to work in the milk federation in 1991–93. Earlier in seventies the Cheshmashahi hills were accessible to the grazing of our livestock. The villagers and village women in large numbers moved to the area for harvesting fodder and carrying firewood for home use. The hill slopes had become bare and did not carry much of area with bush and undergrowth. Despite of having harvesting rights for fodder not much fodder used to be collected from its slopes. However, during the current militancy the area has been cordoned completely from the security point of view for last 17 years. This has completely changed the complexion of the hills around Cheshmashahi. It has now a luxuriant shrub cover all around and so also a rich undergrowth. In fact to reach the Dara Shikoh observatory one finds the footpath leading to it covered with thick foliage and large number of Shah Toot variant of mulberry trees with abundant fruits and so also large crop of bitter cherry fruits growing in abundance. Shahi zeera is also available in abundance. After nightfall the wildlife which had earlier almost become rare prowl around and the chief animals sighted in large numbers are wild black bear, Hangul, Leopard and Jackal. The Cheshmashahi slopes appear densely covered with green cover and in late evening and early morning wildlife prowling around is a common feature and the wild bear made frequent forays in milk parlor of the milk plant and carry milk pouches from the plant by breaking open a window or so which never happened earlier. Similarly large number of Deers (Hangul) have been sighted in late night making their peculiar collective sound a rare phenomena after the exit of Maharaja Hari Singh from the state. During the reign of Maharaja Hari Singh it was a common occurrence to see large herds of Hangul descending in the lower reaches of Dachigam forests to forage on the available herbage and salt licks and feed troughs which were kept for them. This was a golden period for their welfare and protection as the Maharaja was a great lover of wildlife and their protector. Even wild pigs descended in large numbers to the adjoining villages of Harwan and could cause loss to the crops. Villagers would often drive them away with light force. However, wild pigs have been completely eliminated from the wildlife sanctuary and the number of deers considerably reduced. The Cheshmashahi experiment an indirect move to protect the shrubs and forests has reaped a good harvest of luxuriant shrub cover and as a consequence a small sanctuary for the wildlife. Sometimes an errant forceman would get tempted to go for a shikar. The cooperation of army and police has to be ensured so that the wild animals are protected and not disturbed in the created congenial habitat. Kashmir valley is a tourist paradise and the guides are there for trekkers to take them to various places. Trekking routes have to be earmarked as finding new routes and non-serious nature of the picknickers can cause loss of certain rare species of plants and flowers which the visitors can pluck during their trip

Hangul: The Wild Deer of J&K Hills

A luxuriant forest is a peaceful habitat for the wildlife. *Cervus elephus hanglu* is a prized and valued denizen of the valley forests. Presently it is protected in the Dachigam National Park. The park is 22 kms to the north of Srinagar. The area of the park is 22 kms ×

8 kms. It is divided into upper and lower Dachigam. During the Maharaja Hari Singh's regime as stated earlier, the sanctuary contained large numbers of Hangul, with provisions of mangers for feeding them during winter. During winter they would descend on nearly by villages of Harwan. No harm was caused to them and people cooperated in their preservation. In 1947 the number of Hangul in the Rakh was around 2500. In 1959 their number got reduced to 400, in 1966 to 180 and in 1970 to 140. Dachigam was declared to be a Sanctuary in 1951 and Hangul was placed in the list of protected animal species by the Indian Board of Wildlife. Hangul like other wildlife suffered as a result of biotic pressure like grazing, felling of trees for timber and firewood. However, J&K government enforced since 1970 protection of Hangul Habitat strictly and it was so till 1989.However,during the early phases of militancy the enforcement became lax and poaching and felling of trees became an evil practice. However, things have started picking up and enforcement is in place. Recently forest fire in Dachigam caused considerable damage but effective measures were taken and the wildlife protected. However, the shrinking of habitat is forcing the wild animals to descend from mountains to villages in search of food. Recently there have been instances of leopards moving down to villages in various parts of the state and causing loss of human lives and livestock. As a reprisal the villagers took resort to cruel methods of revenge.

The reason for such forage into the human habitations is because of loss of forest area. Such events took place in Tral in Kashmir and Nowshera in Jammu division. Leopards have been sighted at a number of places in valley and Pir-Panjal ranges of Jammu province. There have been instances in recent past of recovering Leopard skins at Lakhanpur post when some Gujjars were caught carrying these for clandestine sale. Leopards have also been sighted in cold arid steppes of Ladakh. There is large scale poaching of these animals which reduces their numbers. Ladakh has the distinction of having a varied stock of wild animals which include wild sheep commonly referred as Shapoo. These animals move in large herds and are large sized with bristly hair. In mid sixties election the author was assigned the duties of Presiding Officer during election time. A police force was with us. A herd of wild sheep was sighted by police force and they fired on an innocent animal. The animal was female and pregnant. The bullet hit its back. The animal got immobilized. It was pregnant. I did my best to treat it. However, its hind legs got paralyzed. It was slaughtered by the police personnel and meat distributed. I heard that many similar instances took place with other parties. This is a sad state of affairs when policemen take law into their own hands to the detriment of the wildlife. Cheru is another prized animal found in the Ladakh and Tibetan steppes. It is a prized antelope whose undergrowth of hair contains famous Shahtoos wool. It has been declared as an endangered animal and trading in shahtoos wool banned. The antelope is presently at the verge of extinction and international effort is on to protect and preserve this species. The other wild animals of Ladakh are wild Yak, Black and brown bear, wild asses and musk deer.

Mussoorie Afforestation Programme and Prevention of Eco-degradation

A unique afforestation programme has been launched in Mussoorie hills and Kumaon hills to improve the forest cover on hill slopes. These slopes had been degraded due to excessive limestone extraction. An infantry battalion of Ex-servicemen of Garhwal Rifles has been created more than a decade back and named as 127 Infantry Battalion and has been provided an honorous job of restoring the grandeur of the green cover on these slopes, which the unit has carried with distinction and discipline. Three million trees have been planted and lush green forests created. This has brought a unique change in the Garhwal hills. The area is now having adequate rainfall and snow which had siezed to occur for decades past has now become a reality in winter. School children are also encouraged to collect tree seedlings and plant them in uncovered areas. The denuded slopes have been restored to their past glory. A similar programme could be launched in the J&K State to restore the forests to past glory and improve the verdant cover of our mountains and hills which have become bare due to reckless felling of trees.

Our Shrinking Wetlands

J&K state has the distinction in having a large number of water bodies in the form of lakes, ponds, springs, bowlies, swamps commonly referred as Nambals and various rivers and rivulets. However, these water bodies are shrinking due to various reasons. The chief reasons being widespread deforestations in the upper catchment areas with the result large scale erosion drives lot of silt and eroded material into these water bodies and reducing the water retention prospects in the higher reach mountains. Large scale siltation has caused the water bodies to become shallow and raised area in the periphery have been encroached by people living in the neighbourhood and hutments and large-scale willows and poplars planted in such raised areas reducing thereby the size of lakes and wetlands.

Wular Lake

It is one of the largest fresh water lake to the north of Srinagar. It is 12 miles long and 8 miles wide. The name wular in Kashmiri means eruption of gushing water from a big hole and so it was earlier named Wular Nag and finally named as Wular. Wular lake in historic times was a bustling city ruled by Sunder Sein. However, this city was submerged due to a severe earthquake and a wide expanse of lake took its present shape. A hundred year back people have spotted big submerged houses and in its eastern part a submerged stone bridge is reported to have been identified. A great temple is also reported to have been located. King Badshah is reported to have raised a garden 100 × 75 yards there. During old times it is reported that the spring in the centre was ejecting water to a height of one yard for considerable period of time. This gushing spring has now receded. In its place bubbles of water now sprout there. Wular is known for its bumper singhara production, varieties of fish, varieties of birds and cattle fodder. The Wular like other water bodies has shrunk due to siltation and considerable pollution. Government is planning to take remedial measures with a massive assistance programme from the central government.

Dal Lake

It is one of the most beautiful lake in Srinagar city spread from Gagribal to Telbal on one side and from Khoj Yarbal to Nishat. It is described as Bod-Dal in front of Hazratbal and Lokut Dal from front of Shankaracharya hill to Nishat and third is in front of Hari-Parbat as Sodar-Khan. Earlier it had an area of 75 sq.kms which at present is reported to have shrunk to 12 sq.kms. Four big springs are reported to be sprouting at four places. In olden times the Dal waters used to join river Jhelum at Habbakadal, but Badshah constructed a Bund and through Nala Mar made provision for carrying Dal waters for irrigation purposes and through Chunti Kul connected it to Jhelum (*Weth*). Nala Mar acted as a irrigational channel moving through the centre of Srinagar for transport and carriage of goods and fodder to various parts of the city. It also acted as a cleansing channel for the water ways so beautifully connected. However, more than two decades back this channel was closed and filled and converted into a wide road connecting various parts of the city. Dal is famous for its fish a large variety of which still thrive in its water. Lotus stem (*Nadru*) and seeds (*Pambach*) are is prized specialty. Various varieties of fodders like *Pech, Har, Nar, Wupai* and *Khor* (*Limnanthemum lymphecoides*) growth extensively on its shores. Khor has been found to be a very nutritious fodder and contains protein as high as 22 per cent on dry matter basis. Pech (marsh grass) is also used for weaving of mats. Various floating gardens are used for growing a variety of vegetables which are sweet and refreshing. However, large scale encroachments and constructions of houses, Hotels and shacks have made the Dal to be a repository of their effluents which is threatening the survival of the lake. The large numbers of houseboats till recent past used to drop their effluents into the lake. This used to be one of the major pollutant of the lake. However, some changes are reported to have been made in the design of the commode installed in the houseboats so that the effluents are now stored instead of being pushed into the lake and then suitably disposed. A Dal Development Authority has been created to relocate the households that had come up in the banks so that pressure of their drainage into the lake is stopped. However, CAG report has castigated lakes and Water Development Authority for its failure to improve the health of Dal Lake despite of incurring an expenditure of 235.70 crores for its improvement. Problems of excessive weed growth, deterioration of water quality, discharge of sewage into the lake body and deposition of the silt still persists, the report said. The open area of the lake has reduced from 24 sq.kms in 1962 to 11.98 sq.km in 1980. Against a target of acquiring 229 hectares of hamlet land and 3.741 structures only 29.5 hectares of land and 528 structure have been acquired till March 2006 the report said. The rate of silt advance into the water body has increased from 7254 tonnes per year (between 8/1998 to 8/2003) to 22,354 tonnes per year between 9/2003 to 8/2005 despite undertaking catchment area management.

Anchar Lake

It lies north of Srinagar city and extends from Omurhair to Biyahama in Ganderbal. During historical times Jayapida is reported to have raised a bund which was known as Mau-Suth. This suth is now almost destroyed and the waters unite with Weth at Shadipora. It is one of the richest wetlands and is visited by lakhs of migratory birds in winter has a sizeable numbers of resident birds. It has large stretches of march grass 'Pechi-Dars', which have wide use as fodder and for production of mats besides other variety of fodders. However, it is at present under great pressure of pollution from Soura Medical Institute which runs hundreds of tonnes of its effluents into this Lake. There was a time when its water was prized for its quality, presently the water is so much polluted that boatmen use special wear to go into its waters as otherwise they catch various skin diseases. The fish catch has got considerably reduced and so is also 'Nadru' disappearing fast.

Khushal Sar

This is a lake to the north of Srinagar, with Zadibal on its east and Idgah to its west. Its waters enter into Anchar Lake. It is also a refuge for large number of birds.

Hokur Sar

It is spread over an area of 75 sq.kms and is 2 miles west of Srinagar and below the Hanjac Wodur. It has large stretches of Pechi-dars and Willow woodlands. It is from ancient times known as as Shikar-Gah for it has a wide range of birds resident and migratory. The other prominent wetlands are Yamasar, near Pattan, Tansar near Village Mansbal. Besldes Haigaṃ and Mirgund also need to be mentioned. These wetlands act as storehouses of water.

Similarly, rivers, ponds and bowlies in Jammu are also under threat and are fast disappearing. About 60 years ago there were 335 big ponds in the Kandi belt of Jammu. These were used for multiple purposes. Most of these have now dried. Sacred streams like *Devika* and *Banganga* are disappearing. The river Tawi in Jammu has become a repository of sewage drains of Jammu city. Protection of the wetlands is a major requirement of the time to stem the tide of destruction of the environment around us in the state as well as the country. This can be achieved by preventing erosion and siltation of the water bodies, Check daming, terracing the hilly slopes, contour trenching and strong measures enforced in smuggling and felling of trees in the forests, strong measures taken for afforestation. Prevention of forest fires which in many cases are engineered by unscruplous elements. Catchment are as are to be guarded of animal trespass and grazing. Effective legislation has to be enacted for treatment of effluents so as not to allow those to be emptied into the lakes and other water bodies. Ladakh has the distinction of having one of the spectacular lake known as Pangong Lake. It is 5 to 6 kms wide and 144 kms long. It lies on Indo-Tibetan border. 40 per cent of its water is in India and rest within Tibet. It is a salt water lake and a remnant of the ice age. It changes its shape depending upon angle one views it. The shades are blue, green, purple turquoise and violet. Its beauty is mesmerizing. Black necked Siberian cranes can be seen swarming in the placid lake making the site enchanting. The herdsmen are also around with their flocks or Pashmina goats and Yaks. Till now the lake was protected from the inflow of tourist. It has now been thrown open for tourists. Great vigil is required to prevent its pollution and protecting its misuse. One of the remarkable glaciers in the Ladakh region is the Siachen Glacier. It is presently being guarded by the Indian troops. There have been apprehensions that the presence of Indian and Pakistani forces in the area may have adverse effect on the recession of the glacier. However, according to Geological Survey of India the recession of the glacier for past 95 years has been 500 metres till year 2004. The annual rate having been 5.3 metres. The recession in other Himalayan glaciers is under scrutiny and this varies from terrain to terrain.

Global Warming and Climate Changes

The World Wide Fund in its project on climate changes has predicted that the largest rivers of the world are dying. Among these are Ganga, Indus Nile, Yangtze, Mekong and Danube. A society cannot countenance the demise of its rivers as these are fundamental to its survival. As such harshest measures have to be taken to preserve our rivers. As regards the climatic changes, the activities in Arctic and Antarctic impact the whole of planet. This is in view of these regions having crucial role in regulating the circulations of the world's oceans, which control the Weather System and climate on every continent. According to the NASAS latest Arctic Climate Assessment report, the temperature in the region of Polar Bear is rising by twice temperature increase of the global average and it could be 10 degree celsius by 2090. The rising green house gases is now melting the planets ice cover.

Ice reflects sunlight back into the space and as it disappears earth absorbs more sunlight which increases warming. This threatens all the ecosystems, rainforests, coral reefs and the ozone shield. Glaciers melt as a consequence at a greater pace. Glaciers store 80 per cent of the worlds freshwater collected during the wet period of the year and slowly releasing it in the drier period. With the faster recession of the glaciers, these may disappear in next hundred years unless effective remedial measures are taken. The loss of arctic ice sheet could spell an eco-disaster and trigger a chain or eco events which could change the face of life on the earth. However, there are contrary views to these gloomy forecasts. Prof. Clark of the Ottawa university, claims that warmer periods of the earth's history came around 800 years before with concurrent higher CO_2 levels. During world war II there was huge rise in CO_2 emissions but the global temperature fell for four decades after 1940. Prof. Philip Scott, emeritus professor of Biogeography opines that climatic change is too complicated to be caused by just one factor. Despite these dissenting voices the general opinion is that global warming is threatening the world ecosystem. A rise in temperature by 2 to 4°F above the current level shall lead to $1/3^{rd}$ or species to be lost from their present range. Coastal wetlands are under threat. Warmer sea waters are bleaching coral reefs, pushing marine species to the poles. Increased CO_2 emissions into the sea water makes the water acidic and inhibits the coral to build skeletons. Polar bears are moving further north. A decline of 20 per cent in their population has been noticed and if the trend of warming continues the polar bear may become extinct in next 100 years. The warming of Asian glaciers is putting at risk tae present availability of 8.6 billion cubic metres of water annually. Since 1970 the average air temperature recorded at 49 stations of the Himalayan region rose by 1°C. Many perennial rivers are likely to turn into seasonal ones. Earlier, it used to snow in the higher reaches of Himalayas in October. Now it snows in later winter. This advances the glacier meltdown. The Gangotri glacier has been receding for past three decades. The Indian Space Research Organization, Ahmedabad has studied 466 glaciers in Chenab, Parbati and Baspa basins through satellite pictures. These glaciers covered 2077 sq.kms in 1962, presently the area has shrunk by 21 per cent.

Sunderbans may Lose to Global Warming

Ghormara Island in Sunderbans has shrunk. During last 30 years nearly 31 sq.miles of Sunderbans has vanished. More than 600 families displaced. Sunderbans are cut by intricate network of streams and tributaries and straddle between India and Bangladesh. 4 million people live in Indian Sunderbans. The rising sea and destruction of forests has threatened the inhabitants who are moving to higher areas for safety. In years ahead both the people who inhabit these islands are threatened, so are the forests. As the earth cracks and submerges, the poor who inhibit the area move further inland.

Livestock and its Role in Eco-degradation

There has been a phenomenal increase in the livestock population in the J&K state as is revealed by the latest livestock population as contained in the statistical digest published by the Directorate of Economics, Statistics, Planning and Development, J&K. While the total livestock population in the state in 1956 comprised 40.08 lakhs, the same has risen to 98.99 lakhs in 2003. The area under grazing lands and pastures has, however, remained, static. The livestock that is almost totally dependant on pastures and grazing is the sheep and goat stock owned by Bakarwals and to some extent by Gaddis. The Bakarwals use the summer pastures of Doda, Poonch and Rajouri and partly some areas of Kashmir valley in summer and shift to Jammu Shivaliks in winter. Gaddis restrict to Bhadarwah in summer and part of Kathua in winter. However, the sheep and goat stock of Gaddis has got considerably reduced due to militancy and some of their dhoks in Kathua have now been taken over by Bakarwals. The inhabitants of foothills of valley are also dependent upon the pastures for grazing their livestock. There is also a procedure in vogue in valley to carry the dry stock to grazing as they otherwise have to be stall fed which is non-renumerative exercise. The following tables give an estimate of the fodder requirements and the area under fodder crops which makes a fair idea of the shortfalls in fodder availability and the pressure it caused on our depleted pasture lands and grazing areas.

The Table 2.1 gives an estimate of fodder requirements in respect of Jammu division. The requirement of Kashmir division is 30 per cent higher than this because of livestock population and severe winter conditions when the livestock has to be stall fed. However, the area under fodder crops is negligible due to importance the cereal and other crops have in the minds of farming community the Table 2.2 gives such details as to the availability of the area under fodder crops.

Table 2.1

Sl.No.	*Category of Livestock*	*Population as per 2004–05 Census in Lacs*	*Fodder Requirement in Lac Metric Tonnes*
1.	Cattle	17.111	25.67
2.	Buffaloes	10.203	15.30
3.	Sheep	22.58	8.24
4.	Goat	14.67	5.36
5.	Others	1.485	2.24
	Total	66.06	56.81

Table 2.2

Sl.No.	*District*	*Total Area*	*Net Area Sown*	*Total Area Sown*	*Area Under Fodder Crops*	*Percentage Area Under Fodder Crops*
			(Lac Hectares)			
1.	Jammu	3.21	1.190	2.13	0.150	7.04
2.	Udhampur	4.32	0.697	1.12	0.0042	0.37
3.	Doda	4.11	0.693	0.83	0.0072	0.86
4.	Kathua	2.65	0.638	1.25	0.076	6.08
5.	Rajouri	2.52	0.529	0.98	0.009	1.01
6.	Poonch	1.14	0.273	0.46	0.231	5.88

Despite of the appreciable increase in the livestock population the per cent area under fodder crops in Jammu division is 3.98 per cent only. This is a dismal picture lending pressure on the grazing lands and pastures. Since sheep and goat are by far the most stock put loose on natural forest land pastures the damage to the herbs is most glaring and resulting in the extinction of the most herbal cover and putting much pressure on the shrubs by the goats by virtue of their grazing habit and harvesting these for the sheep with long poles with a metal contrivance in place by the bakarwals for denuding the palatable leafy trees. The damage is clossal. Table 2.3 gives the Division-wise statistics of pasture/range lands within demarcated forests.

Table 2.3

Sl.No.	Division	Geographical Area	Forest Area	Percentage of Forest Area	Pasture and Grazing Land		
		In Lac Hectares			Alpine	Others	Total
1.	Kashmir	15. 948	8.115	50.80	2.102	0.433	2.535
2.	Jammu	20.293	12.050	45.83	0.928	0.730	1.658
3.	Ladakh	90.701	0.017	0.02	–	–	–
	Total	138.942	20.182	14.52	3.030	1.163	4.193

The production and carrying capacity is extremely reduced due to overgrazing and an increased livestock population with the Bakarawal sheep and goat stock completely dependent upon these depleted grazing lands and as a consequence causing widespread degradation of the herbal and shrub cover of our forests. In the Table 2.4 area available in hectares per capita of livestock is given to present the grim situation it bodes for the livestock and regeneration of our forests.

Table 2.4

Sl.No.	Province	Area per Capita Livestock in Hectares			Percentage Decrease from 1977
		1977	1997	2004–05	
1.	Kashmir	0.143	0.069	0.076	47%
2.	Jammu	0.064	0.030	0.025	61%
	Total	0.090	0.046	0.042	53%

Consequent upon the increase in livestock population the pressure on pasture and grazing lands has increased manifolds. In records there is a sizeable 16.67 lac hectares shown as barren lands, wastelands and roadside lands but this is susceptible to land grab position of the influential people and with new Roshini Act in place, the status of this land is suspect and not available for conversion to

Table 2.5: Sheep and Goat Population of Sheep Husbandry Department Jammu as per Livestock Census 2003

Sl.No.	*District*	*Sheep Population*			*Goat Population*		
		Cross-breed	*Local*	*Total*	*Male*	*Female*	*Total*
1.	Jammu	0.73	0.16	0.89	0.44	1.43	1.87
2.	Kathua	1.31	0.97	2.28	0.44	1.40	1.84
3.	Udhampur	3.46	2.30	5.76	0.84	3.27	4.11
4.	Doda	3.34	3.43	6.77	0.50	1.46	1.96
5.	Rajouri	2.36	1.73	4.09	0.80	2.46	3.26
6.	Poonch	1.76	1.01	2.77	0.45	1.19	1.64
	Total	12.96	9.60	22.56	3.47	11.21	14.68

Table 2.6: Sheep and Goat Population of Sheep Husbandry Department Kashmir as per Livestock Census 2003

Sl.No.	*District*	*Sheep Population*			*Goat Population*		
		Cross-breed	*Local*	*Total*	*Male*	*Female*	*Total*
1.	Anantnag	1.62	0.88	2.50	0.15	0.36	0.51
2.	Baramulla	1.75	1.15	2.90	0.25	0.71	0.96
3.	Badgam	0.81	0.24	1.05	0.13	0.28	0.41
4.	Kargil	0.27	0.67	0.94	0.24	0.52	0.76
5.	Kupwara	0.23.	0.53	0.76	0.08	0.35	0.43
6.	Leh	0.55	0.47	1.02	0.75	1.44	2.19
7.	Pulwama	1.12	0.25	1.37	0.11	0.24	0.35
8.	Srinagar	0.69	0.30	0.99	0.07	0.18	0.25
	Total	7.04	4.49	11.53	1.78	4.08	5.86
	Grand Total	20.00	14.09	34.09			20.54

Source: Sheep Husbandry Department, Jammu.

pasture lands. The only silver lining in the livestock development programme has been the production of crossbred jerseys and Holstein Friesian animals which are high yielders of milk and as a consequence stall fed as no owner can afford to put such heavy milkers on pasture and field grazing. A national policy has to be put in place to determine alternate plans for the Bakarwal community who are solely dependent upon the forest grazing lands for feeding their sheep and goat stock. This can involve aerial fertilization of such forest grazing lands in phases and closing such fertilized areas for such time till they have a luxuriant herb cover and shrubage. A coordinated approach has to be arrived to protect the forests from too burdensome animal load for grazing so as to protect them in the best interests of environment.

Summary

The eco-degradation in J and K state, the country as a whole is fast approaching a stage when the life and living is threatened by shrinking water bodies and wetlands, widespread pollution of air, receding glaciers and global warming. Climate changes in J&K are being increasingly experienced as happened in February 2005 when heavy snowfall occurred and was referred as Snow Tsunami effect. The unprecedented rains in Ladakh during August 2006 caused floods, cloud bursts and loss or residential houses. The increased number of animals dependent on pasture grazing in forests is reducing the herb and shrub cover of our forests. The whole gamut of factors in the cause of development of industry and agriculture has put pressure on our forest cover and threatened the extinction of our wildlife. Increase in the livestock is emerging as a cause of green house gases leading to global warming. All these varied causes of degrading our environment have been appraised and a word of caution sounded to protect our resources whether it be our water bodies, forests, glaciers wildlife and our surroundings for a better life to us and future generations.

References

Ahmed, Shamasdin, 1998. *Tarikhi Hassan,* Kashmiri Translation. J&K Academy of Art and Culture, Srinagar.

Anonymous, 13-3-07. Centre to define forests afresh. *Times of India,* New Delhi.

Anonymous, 13-3-07. Snow good for water Bodies. *Hindustan Times* Chandigarh.

Anonymous, 2004–2005. J&K Government, Digest of Statistics, Directorate of Economics and Statistics, Planning and Development Department.

Anonymous, 3/2007. Climate change threaten world ecosystem. *Times of India*, New Delhi.

Gupta, R.D., 3/2007. Shrinking wetlands. *Daily Excelsior*, Jammu.

Kaul, V.K. and Pandit, A.K., 1979. Is Dal Lake dying. *Science Reporter*.

Kaw, A.K., 2006. Project proposals for fodder production in pasture lands, Jammu province. Directorate of Sheep Husbandry, Jammu.

Mahapatra, Dhananjay and Sethi, Nitni, 8-3-07. Filth threatens source of Ganga. *Times of India*, New Delhi.

Rahmani, Asad Rafi, 1979. Our threatened plant species. *Science Reporter*.

Singh, S.D. Jamwal, 3/2007. Stunning, mesmerizing Pangong Lake. *Daily Excelsior*, 11.3.2007.

Tiku, J.L., 1992. Cattle development in J&K Himalayas. In: *Himalayan Environment Man and Economic Activities*, (Ed.) J.L. Raina. Pointer Publishers, Jaipur, 2: 459–482.

Tiku, J.L., 1995. Pastoral livestock and eco-degradation in Jammu and Kashmir. In: *Eco-degradation of Himalayas: Causes and Remedies*, (Ed.) Bansi Lal Kaul, Vinod Publishers, Jammu.

Chapter 3

Characteristics, Genesis and Classification of Some Soils from Sub-mountain Region of Jammu Siwaliks

R.D. Gupta

ABSTRACT

Seventeen soil profiles representing major soils in varying physiographic stretch and geological formations, across the Siwalik hills of Jammu region, Jammu and Kashmir State were studied for their characteristics, genesis and classification. About 50 per cent of soil profiles were characterised by the presence of more content of gravels and pebbles which make them difficult to tillage. All soils are shallow in depth (<1m). They exhibit AB and AC profiles depending upon the topography, parent material and vegetation. The soils are derived from Siwalik group of rocks belonging to Middle Miocene, Pliocene and lower Pleistocene period, younger in age which seem to have dominant role in the genesis of these soils. The other pedogenic factors influencing their formation are topography and vegetation. With time, however, the effect of parent material seems to be obliterated.

On the whole, the soils formed on mild/gentle slopes were found to exhibit the development of altered structural B (cambic) horizon. These, therefore, have been classified either into Typic Ustochrepts and Fluventic Ustochrepts or Typic

Eutrochrepts and Dystric Eutrochrepts. The soils showing no horizon differentiation are classified into Typic Ustorthents and Aquic Ustorthents.

Introduction

The Siwalik hills of Jammu region in the Jammu and Kashmir, are spread in four districts *viz.*, Jammu, Kathua, Udhampur and Rajouri. Despite having very high potentials in soils of these hills, a number of constraints/problems such as undulating topography, water erosion of various kinds including gully and ravine formation, denudation, deforestation and droughtness limit their productivity for agricultural crops. There is scanty information available with regard to the characterisation of these soils which is essential for suggesting appropriate management practices and alternate land use system for sustained production. Attempts are, therefore, made in the present study to throw some light on morphology and physico-chemical characteristics of these soils, and ameliorating measures thereon. Genesis of these soils has also been detailed vis-à-vis their classification according to soil taxonomy.

Geographical Conditions

Physiography

The Siwalik sub-mountain hills in the Jammu region constitute the *Kandi* belt. These are also called the Southern most Himalayas or Sub-Himalayas. This belt, stretches between 74°21′ to 75°52′ E and 32°22′ to 32°55′ N, starting from Basohli and traversing the three districts *viz.*, Kathua, Jammu and Udhampur and ending upto the lower hills of Rajouri district. The altitude of these hills varies from 300 to 1050 m above mean sea level (Gupta and Banerjee, 1991).

Drainage

The area is drained by a large number of seasonal rivulets or streams locally called *Choes* or *Khuds* as well as perennials. Among the perennials, these are the Jhajar, Balini, Basantar, Tawi, Seot, Manawar Tawi and Ujh. The first six constitute the tributaries of the river Chenab and seventh one that of the river Ravi.

Geology

The base of the Siwaliks is made up of alernating beds of hard yellowish green sandstones to red colour shales. These beds regularly alternate with a distinct pattern. There are distinct ripple marks on the top of beds indicating the presence of slow and steady rivers responsible for their formation. These so called Lower Siwalik rocks, are seen on a road side on National Highway between Nandni and Jhajar Kotli.

The alternating shale-sandstone sequence is succeeded by thick sandstone deposits of characteristics with white and pepper colour. The sandstones contain a lot of cementing materials which resulted in hard lenticular layers within the sandstone. These broken parts of lenticles have produced pieces of the sandstones having varied form and shape like round, cliptical and ball shaped. These types of rocks are called Middle Siwaliks. The sandstones of these rocks have few beds of clay or shale. The Middle Siwalik rocks are found in the vicinity of Ban Talab, as well as Chouki Chora block of Akhnoor. These rocks are also found in Bhadu Bilawar and Basohli tehsils of Kathua district.

Middle Siwalik sandstones are succeeded by sandstone-shales squence constituting the Upper Siwaliks. These rocks however, are of lighter colour, pink and lighter coloured sandstones with lenticles of coarse material of cobble and shingle. These are found in Basohli tehsil of Kathua as well as parts of Kathua and Rajouri.

Climate

Generally the climate of whole of the Jammu sub-mountain Siwaliks is characterised by sub-tropical nature. The rain-fall varies from 1000 to 1340 mm annually with an average of 1150 mm, the bulk of which occurs during monsoon *i.e.*, from July to September. The maximum temperature usually remains 38–40°C, but frequently goes upto 42–44°C during June. January is the coldest month of the year with maximum and minimum temperatures of 19.5 and 6.8°C. June is the hottest month with average maximum and minimum temperatures of 38.9 and 28.7°C.

Vegetation

The natural vegetation of the sub-mountain Siwaliks mostly consists of subtropical dry mixed forests of scrub type. The forests of *Kandi* belt have been classified into five forest groups *viz. Acacia, Bauhinia, Dodonaea,* Mixed semi-deciduous and chir forests. *Acacia nilotica, A. modesta, A. catechu, Grewia optiva, Butea monosperma,*

Dalbergia sisso are the main tree species alongwith *Pinus roxburghii, Adhatoda vasica, Carissa spinarum, Cannabis sativa, Dodonaea viscosa* and *Vitex negundo* are the main shrubs.

Materials and Methods

Based on reconnaissance soil survey, seventeen soil profiles were selected for the purpose. Morphology of the soils were studied in the field and classified taxonomically. Physico-chemical properties were determined following standard procedures.

Physico-Chemical Properties

Mechanical Composition

Presence of more sand in surface horizon in most of the soils might be due to surface erosion and/or clay translocation (Table 3.1). It is further observed that soils of profiles 1,3–5, 8, 10 and 11 have relatively higher content of coarse sand ranging from 15 to 30 per cent as compared to others which is attributed to richness of this fraction in the parent materials from which these soils are formed. It is also remarkable to note that in soils of profiles 1 to 8 and 11, there was more content of gravels/pebbles in the surface to the extent of 0.1 to 3.0 per cent by volume or some times even more. According to the limits, the soils of these profiles belong to St_2 stoniness category (Kaul, 1974) which make these soils difficult to till. Although translocation of clay has been observed in B or C horizons of profiles 1, 5 to 6, 10 to 12 and 13 to 17 yet none of the horizons of these profiles qualify for an argillic one. This was further supported by the absence of clay skins or clay cutans on the ped faces of these horizons.

pH, $CaCO_3$ and Organic Carbon Contents, CEC and Exchangeable Cations

The pH of the soils varies from 7.1 to 8.8 irrespective of soil depth. There is general leaching of $CaCO_3$ in soil of 2,8,10 and 14 to 17 profiles which is evident from increase in pH with depth (Table 3.1). In soils of other profiles though free $CaCO_3$ was present yet it lies within possible experimental error of less than one per cent and hence has been neglected (Gupta, 1992). The calcareous nature of the soils can be ascribed to its inheritance from the parent material.

Almost all soils have low amount of organic carbon (<1 per cent) which usually decreases with depth. Its low amount is

Table 3.1: Mechanical Composition and Physico-chemical Properties of Soils from Siwalik Hills of Jammu

Mechanical Composition (Per cent)							Physico-chemical Properties							
Depth (cm)	Horizon	Clay	Silt	Fine Sand	Coarse Sand	Texture	pH	$CaCO_3$ (%)	Organic Carbon (%)	CEC and Exchangebale Bases				
										CEC (me/100 g)	Ca	Mg	K	Na
Profile 1 (Village Chirore, elevation 320 m, slope 1 per cent, sandstone intercalated with pink coloured clay, *Dalbergia sisoo, Acacia nilotica*)														
0–9	A1	14.1	14.0	41.3	30.0	S1	7.3	–	0.75	7.8	5.5	1.4	0.4	0.1
9–27	B1	17.0	15.5	50.1	19.0	S1	7.1	–	0.60	7.3	5.6	0.8	0.6	0.2
27–41	B21	14.9	14.0	47.0	22.0	S1	7.2	–	0.30	6.1	3.8	1.9	0.7	0.1
41–81	C	16.3	17.0	59.0	6.0	S1	7.2	–	0.15	4.3	2.3	1.2	0.4	0.1
Profile 2 (Ram Nagar, elevation 310m, slope 3–5 per cent, soft sandstone calcareous, *Pinus roxburghii, Dodonaea viscosa, Carissa spinarum*)														
0–10	A1	18.4	31.2	43.9	10.0	Sil	7.6	–	1.20	10.2	6.6	2.1	1.1	0.3
10–21	A2	18.9	32.8	40.4	8.0	Sil	8.0	–	1.05	9.8	5.6	1.7	1.7	0.4
21–61	B1	16.4	40.5	40.0	4.0	Sil	8.0	–	0.15	7.1	4.3	0.6	0.6	0.6
Profile 3 (Balwal, elevation, 311m, slope 4–5 per cent, sandstone, *Acacia nilotica, Acacia modesta*)														
0–13	A1	14.1	15.6	57.8	13.8	S1	7.7	–	0.57	10.8	5.1	4.1	0.1	0.1
13–26	A2	14.8	14.5	56.8	14.8	S1	7.5	–	0.30	10.2	5.7	3.3	0.1	0.1
26–57	B1	14.7	13.0	55.0	15.9	S1	7.4	–	0.22	8.6	5.0	3.0	0.1	0.1
57+ pebbles stones were found														

Contd...

Table 3.1–Contd..

Depth (cm)	Horizon	Clay	Silt	Fine Sand	Coarse Sand	Texture	pH	$CaCO_3$ (%)	Organic Carbon (%)	CEC	Ca	Mg	K	Na
	Mechanical Composition (Per cent)						Physico-chemical Properties			CEC and Exchangebale Bases (me/100 g)				
Profile 4 (Raipur, elevation 320m, slope 3–4 per cent, sandstone interrelated with clay, *Acacia niloctica, Adhatoda vasica*)														
0–17	Ap	17.5	13.4	52.5	18.1	81	7.8	–	0.62	8.7	5.8	1.2	0.2	0.3
17–29	B1	14.4	14.8	48.7	17.9	1	7.9	–	0.51	8.4	6.3	0.5	0.6	0.2
29–75	B12	15.2	20.1	44.9	21.5	1	7.4	–	0.22	6.4	3.8	1.6	1.4	0.3
75–91	C1	13.0	20.0	49.9	18.1	1	7.6	–	0.20	6.0	3.0	1.2	1.4	0.4
Profile 5 (Village Kote, elevation 330 m, slope 1–2 per cent, silt stone/hard sand stone, uncultivated *Albizia lebbek*)														
0–21	Al	18.4	26.8	36.7	18.5	Sil	7.6	–	0.45	8.7	4.1	1.6	1.3	0.4
21–57	C1	22.0	19.0	39.8	20.4	I	7.5	–	0.27	7.3	4.8	0.5	0.6	0.5
57–86	C2	19.0	21.4	44.0	15.7	I	7.7	–	0.21	7.4	5.9	1.4	0.3	0.2
86+ Pebbles and stones were found after 2nd layer														
Profile 6 (Amb, elevation 360m, slope 1–2 per cent, sandstone, *Acacia species*)														
0–22	A1	24.0	37.8	29.8	9.8	Sil	7.5	–	0.27	8.1	5.1	0.8	0.9	0.7
22–47	C1	27.0	35.6	30.6	7.0	Sil	7.8	–	0.35	7.3	4.9	1.5	0.9	0.1
47–71	C2	20.0	39.5	37.2	4.1	Sil	7.9	–	0.06	6.1	3.5	1.3	0.8	0.4

Contd...

Table 3.1–Contd..

Mechanical Composition (Per cent)							Physico-chemical Properties								
Depth (cm)	Horizon	Clay	Silt	Fine Sand	Coarse Sand	Texture	pH	$CaCO_3$ (%)	Organic Carbon (%)	CEC and Exchangebale Bases					
										CEC	Ca	Mg	K	Na	
										(me/100 g)					
Profile 7 (Gandarwan, elevation 490 m. slope 1–4 per cent slope, alluvium, eroded land, scrub vegetation)															
0–27	Ap	16.0	12.0	62.1	12.1	S1	7.6	–	0.30	7.7	4.8	2.1	0.4	0.1	
27–61	C1	13.0	12.0	60.0	11.1	Sil	7.5	–	0.22	6.6	3.1	1.4	0.2	0.1	
57–86	C2	13.0	14 0	64.0	13.0	Sil	7.4	–	0.15	5.9	3.0	2.0	0.2	0.1	
Profile 8 (Hill forest near, Bahu village, elevation 400m, slope 4–5 per cent, sandstone/conglomerate, *Acaca nilotica modesta*)															
0–16	A1	16.0	14.0	50.1	20.0	S1	7.5	1.1	0.87	11.4	6.1	3.8	0.2	0.1	
16–27	B12	14.6	15.0	52.2	18.2	S1	7.8	1.4	0.25	9.4	4.9	3.0	0.2	0.1	
27–61	B2	13.0	16.0	53.6	17.9	S1	7.9	1.5	0.15	8.0	4.8	2.7	0.2	0.1	
61–81	C1	14.0	17.0	49.8	21.0	S1	7.6	1.1	0.07	5.4	3.9	2.0	0.2	0.1	
Profile 9 (Uttar Bani, elevation 425m, slope 1–3 per cent, Gheomitti, clays, *Acacis* species)															
0–23	A1	36.9	14.1	42.3	6.4	cl	7.9	–	0.89	12.6	10.0	1.3	0.1	0.3	
23–37	B1	22.8	16.6	56.0	4.6	l	8.1	–	0.64	15.2	12.6	0.9	0.9	0.3	
37–71	B2	25.6	19.4	47.0	5.9	l	8.1	–	0.58	10.9	8.9	0.8	0.8	0.2	

Contd...

Table 3.1–Contd..

Mechanical Composition (Per cent)							Physico-chemical Properties							
Depth (cm)	Horizon	Clay	Silt	Fine Sand	Coarse Sand	Texture	pH	$CaCO_3$ (%)	Organic Carbon (%)	CEC and Exchangebale Bases				
										CEC (me/100 g)	Ca	Mg	K	Na
Profile 10 (Atham, elevation 500m, slope 1–2 per cent, sandstone intercalated with clay, calcareous, *Pinus roxburghii* forest)														
0–19	Ap	22.0	20.8	48.4	13.8	I	7.5	2.2	0.65	8.8	5.7	1.2	1.2	0.4
19–37	11B1	25.4	15.4	35.9	15.7	Cl	7.7	2.5	0.49	7.6	5.2	0.9	0.9	0.3
37–56	11B121	17.0	19.0	39.0	24.0	I	8.0	2.6	0.40	9.4	6.5	1.2	0.9	0.4
56–89	11B22	21.0	24.0	44.0	11.7	I	8.4	2.9	0.24	6.1	4.0	0.8	0.9	0.2
Profile 11 (Bara, elevation 320m, slope 4.5 per cent, undulating, sandstone/conglomerate, *Dalbergia sisso*)														
0–26	Al	17.6	18.8	46.4	17.8	I	7.4	–	0.61	8.0	4.8	1.7	0.9	0.5
26–47	B1	23.7	14.7	42.6	21.9	I	7.4	–	0.07	4.8	2.7	0.9	0.4	0.2
47–48	B2	19.4	17.9	41.4	21.4	I	7.6	–	0.15	8.2	5.1	1.2	1.5	0.4
Profile 12 (Sangar, elevation 553m, slope 1–2 per cent, sandstone mixed with mudstone, Govt. forest land)														
0–17	A1	20.1	16.1	59.9	1.2	I	7.7	–	0.60	13.2	8.1	4.1	0.2	0.1
17–36	B21	23.1	15.1	55.0	2.2	I	7.7	–	0.30	12.9	6.9	5.4	0.2	0.1
36–58	B22	24.1	14.0	57.0	3.1	I	7.9	–	0.27	11.9	6.1	4.9	0.2	0.1
58–81	B23	19.0	15.0	58.0	4.1	I	7.8	–	0.07	9.1	5.8	2.2	0.3	0.2

Contd...

Table 3.1–Contd..

Mechanical Composition (Per cent)							Physico-chemical Properties							
										CEC and Exchangebale Bases				
Depth (cm)	Horizon	Clay	Silt	Fine Sand	Coarse Sand	Texture	pH	$CaCO_3$ (%)	Organic Carbon (%)	CEC	Ca	Mg	K	Na
										(me/100 g)				
Profile 13 (Jandra, elevation 740 m, sandstone mixed with clays, mixed forests)														
0–21	A1	21.4	18.9	56.0	4.0	1	7.8	–	0.61	11.2	7.0	2.8	0.2	0.1
21–41	B1	22.4	16.0	51.0	5.0	1	7.9	–	0.62	10.9	10.9	3.0	0.2	0.1
41–66	B2	24.8	17.0	53.0	6.0	1	7.4	–	0.22	8.5	8.5	1.1	0.2	0.1
6+ A layer of sandstone/gravels														
Profile 14 (Mansar, elevation 710 m, slope 1–2 per cent, sandstone mixed with shales, *Pinus roxburghii*)														
0–22	A1	13.0	12.8	60.0	14.0	S1	7.9	1.8	0.75	14.9	7.1	5.9	0.3	0.1
22–45	B1	19.0	20.0	57.0	6.7	l	8.0	2.3	0.45	11.4	8.0	3.2	0.2	0.1
45–73	B12/C	14.0	40.0	40.8	5.1	Sil	8.0	2.2	0.30	8.0	5.0	2.2	0.3	0.1
Profile 15 (Surinsar, elevation 585 m, slope 1–2 per cent, pink coloured clays mixed with sandstone, *Pinus roxburghii*)														
0–11	Al	20.0	34.0	42.0	1.1	Sil	7.5	1.4	0.69	13.4	7.5	4.5	0.2	0.2
11–27	B1	26.0	31.8	42.0	1.0	Sil	8.8	1.3	0.55	12.9	6.9	5.4	0.3	0.4
27–41	B12	20.9	35.2	40.0	3.8	Sil	8.0	2.1	0.60	11.8	6.0	5.4	0.2	0.2
41–64	B22	24.0	34.8	41.0	2.1	Sil	8.2	4.0	0.24	10.4	5.0	4.5	0.2	0.3
64+ A layer of clay intermixed with shale/sand stone														

Contd...

Table 3.1–Contd..

Mechanical Composition (Per cent)							Physico-chemical Properties							
Depth (cm)	Horizon	Clay	Silt	Fine Sand	Coarse Sand	Texture	pH	$CaCO_3$ (%)	Organic Carbon (%)	CEC and Exchangebale Bases				
										CEC (me/100 g)	Ca	Mg	K	Na
Profile 16 (Surinsar, elevation 585 m, slope 2–3 per cent, shale intermixed with sandstone, *Dodonae viscosa*, forest)														
0–14	Al	20.1	34.0	42.0	1.1	Sil	7.5	1.4	0.69	13.4	7.5	4.5	0.2	0.2
14–29	A2	22.0	31.8	42.0	1.0	Sil	8.8	1.3	0.55	12.9	6.9	5.4	0.3	0.4
29–73	B1	24.0	35.2	40.0	3.8	Sil	8.0	2.1	0.60	11.8	6.0	5.4	0.2	0.2
73–91	B22	26.0	34.8	41.0	2.1	Sil	8.2	4.0	0.24	10.4	5.0	4.5	0.2	0.3
Profile 17 (Seot. elevation 700. slope 1 per cent, sandstone/clay, *Dodonaea viscosa*														
0–11	A1	16.1	13.4	60.1	10.1	Sl	7.6	1.4	0.60	10.8	7.2	2.4	0.8	0.1
11–22	A2	19.4	12.6	61.2	9.9	Sl	7.7	1.7	0.69	11.0	3.8	4.5	0.6	0.1
22–44	B21	20.0	14.1	67.0	6.2	Sl	7.8	2.0	0.25	9.5	4.6	3.6	0.7	0.2
44–71	B22	21.2	13.1	66.0	8.0	Sl	7.9	3.1	0.47	8.8	2.2	4.8	0.4	0.1

71* A layer of sandstones/gravels

explained due to the existing hyper thermic temperature (>22.6°C) responsible for hastening organic matter oxidation vis-à-vis scarce vegetation of the area and eroded nature of the soils. Similar observations have been reported in the past (Nazar, 1993 and Gupta, 1994). It is pertinent to mention that in soil profiles of 10, 13, 15 and 17, the distribution of organic carbon is quite uniform even upto the depth of 40 cm or more (Table 3.1) after which it abruptly falls down despite erosion hazards. It is suggestive of coagulation of humus particles by Ca^{2+} ions which prevent their dispersion and subsequent removal as well as due to high degree of protection by clay particles.

Almost all soils have low values of CEC ranging from 5.4 to 15.2 me $100g^{-1}$ with dominance of Ca^{2+} followed by Mg^{2+}. The values of CEC appear to be influenced by organic carbon and clay contents.

Morphological Characters

Morphological characteristics (Table 3.2) reveal that these soils exhibit AB or AC profiles depending upon the topography, parent material and natural vegetation. Barring 7, all profiles are shallow in depth *i.e.*, less than 1 m, but it is more in case of profiles 3, 11–12, 15 and 17 where the depth is upto 60 or 80 cm.

Surface soils are mostly characterised by brown (10 YR5/3, 10 YR5/4), dark brown (7.5 YR4/3, 7.5 YR3/2) and reddish brown in colour (5 YR5/4, 2.5 YR6/4) and those of subsurface by reddish brown (5 YR5/4), dark brown (7.5 YR3/2) brown (10 YR5/3) and greyish dark, greyish brown (10 YR5/2, 2.5 YR3/2). This is suggestive of more or less well drained conditions of the studied area. However, yellowish brown colour in soils of profiles 5 and 14 indicates poor drainage. The reddish brown matrix colour of Al and Ap horizons of profiles 2 and 4 and Al that of 9 is attributed to more presence of ferruginous minerals in the form of magnetite and haematite responsible to impart reddish colour to the soils (Gupta and Verma, 1992). There is lot of textural variation in the soils (Table 3.2) due to heterogeneous nature of parent material and topographical differences. Similar textural heterogenity has also been found earlier (Gupta, 1994). Subangular blocky to blocky structure in soil profiles of 1, 5–6, 9, 11–14 and 15–17 is indicative of weak to moderate profile development. The permeability of the soils is rapid to slow depending upon the clay content (Table 3.2).

Table 3.2: Morphological Characteristics of Soils from Siwaliks of Jammu Region

Profile	Horizon	Colour	Texture	Structure	Permeability	Reaction with HCl	Consistence	Special Characters
1.	A1	10YR 5/3	sl	wcgr	ra	–	fr	Gravels and pebbles
	B1	5YR 5/4	sl	wcgr to lf	ra to m	nil	fr	–
	B21	5YR 5/4	sl	1 fgr	ra	nil	fr	–
	C	10YR5/2	sl	3c abk	ra	nil	mfr	A layer of sandstone was found after 61 cm depth
2.	A1	2.5YR 5/4	sil	1 fwgr	ra to m	e	fr	Gravels and pebbles
	A2	10YR 5/2	sil	2mwgr	ra to m	es	fr	–
	B1	10YR5/2	sil	1 fwgr	ra	e	fr	–
3.	A1	10YR 5/3	sl	lf wc gr	ra	nil	fr	Gravels and pebbles
	A2	5YR 5/4	sl	lf wc gr	ra	nil	fr	–
	B1	5YR 5/4	sl	lf wc gr	ra	nil	fr	–
4.	Ap	2.5YR 6/4	sl	lfc gr	ra	nil	fr	Gravels and pebbles
	B1	10YR 6/2	l	1 f mgr	ra to m	nil	fr	–
	B12	10YR 5/2	l	1 f mgr	ra to m	nil	fr	–
	Cl	10YR 5/2	1	1 f mgr	ra to m	nil	fr	A layer of sandstone

Contd...

Table 3.2–Contd...

Profile	Horizon	Colour	Texture	Structure	Permeability	Reaction with HCl	Consistence	Special Characters
5.	Ap	10YR 5/4	sil	wcgr	ra	nil	fr	Gravels and pebbles
	CI	10YR 6/3	sil	wlf gr to 3c abk	ra to m	nil	mfr	were found throughout
	CII	7.5YR 6/3	sil	wlf gr to 3 abk	ra to m	nil	fr	the profile
6.	Ap	7.5YR 4/3	sil	wlf gr to wc gr	ra	nil	fr	Gravels and pebbles
	CI	7.5YR 5/6	sil	3c abk to bk	ra to m	nil	mfr	were found throughout
	C11	7.5YR 5/6	sil	wl f gr to we g	ra to m	nil	fr	the profile
7.	Ap	7.5YR 4/3	sl	wc gr	ra	nil	fr	Gravels and pebbles
	C1	7.5YR 5/6	sl	wc gr	ra	nil	fr	were found throughout
	C11	7.5YR 5/6	sl	wc gr	ra	nil	fr	the profile
8.	A1	10YR 5/3	sl	lf gr	ra	e	fr	
	C12	7.5YR 5/4	sl	lf gr	ra	e	fr	
	B2	10YR 4/2	sl	lf gr	ra	es	fr	
	C1	10YR 4/2	sl	lf gr	ra tos	es	fr	A layer of sandstone was found
9.	A1	5YR 5/4	cl	lf gr to sbk	ra to s	nil	mfr	
	B1	7.5YR 3/2	1	lf gr to abk	ra to s	nil	mfr	
	B2	10YR 3/2	1	lf gr to abk	ra to s	nil	mfr	

Contd...

Table 3.2–Contd...

Profile	Horizon	Colour	Texture	Structure	Permeability	Reaction with HCl	Consistence	Special Characters
10.	A1	7.5YR 3/2	1	lf gr to wf gr	ra	e	mfr	
	11B 1	7.5YR 5/2	cl	1 f sbk	ra to s	e	mfr to h	
	11B21	10YR 5/2	1	lf sbk to abk	ra	e	mfr	
	11B22	10YR 4/2	1	2m sbk to abk	ra	e	mfr	
11.	A1	10YR 8/3	1	lf gr to wcgr	ra	nil	fr	Pebbles and gravels
	B1	2.5YR 3/2	1	lf gr to sbk	ra	nil	fr to mfr	
	B2	2.5YR 3/2	1	2m sbk to abk	ra to s	nil	mfr	
12.	A1	10YR 5/4	1	1 fw gr	ra	nil	fr	
	B1	10YR 5/4	1	lf gr to 2m sbk	ra to s	nil	fr to mfr	
	B22	10YR 5/4	1	lfw gr to 2m sbk	ra to s	nil	mfr	
13.	A1	7.5YR 3/4	l	lf gr to 2m sbk	ra to s	nil	mfr	
	B1	7.5YR 3/4	l	lf gr to 2m abk	ra to s	nil	mfr	
	B12	7.5YR 3/2	l	lf gr to 2m abk	ra to s	nil	mfr	
14.	A1	10YR 5/3	l	lf wgr	ra	e	mfr	
	B1	10YR 5/4	l	lf wgr to 2m sbk	ra to s	e	mfr	
	B12/C	7.5YR 4/3	l	lf wgr	ra	e	mfr	

Contd...

Table 3.2–Contd...

Profile	Horizon	Colour	Texture	Structure	Permeability	Reaction with HCl	Consistence	Special Characters
15.	A1	10YR 5/3	sil	lf wgr	ra	e	mfr	
	B1	5YR 5/4	sil	lfw gr to 2 sbk	ra to s	e	mfr	
	B12	7.5YR 5/4	sil	2m sbk	ra to s	e	mfr	
	B22	7.5YR 5/4	sil	2m sbk	ra to s	e	mfr	
16.	A1	7.5YR 5/4	sil	lf wgr	ra	e	mfr	
	A2	10YR 5/3	sil	lfw gr to 2m sbk	ra to s	e	fr to mfr	
	B1	10YR 5/3	sil	2m sbk to 3 c abk	ra to s	e	mfr	
	B12	10YR 5/3	sil	sbk to abk	ra to vs	e	mfr	
17.	A1	5YR 5/4	sl	lf wgr	ra	e	fr	
	B1	7.5YR 5/4	sl	lfw gr to 2m sbk	ra to s	e	fr	
	B2	7.5YR 5/4	sl	l fwgr to 2m sbk	ra tos	e	fr	
	B22	7.5YR 5/4	sl	l fwgr to 2m sbk	ra to s	e	fr	

Note: Symbols followed are those of Soil Survery Manual (Soil Survey Staff, 1951).

Genesis and Classification of Soils

Genesis

Parent material, topography and vegetation seem to be the main pedogenic factors for the genesis of the soils of present study.

The nature of parent material, markedly influences the soil characteristics in the initial stages. Robinson (1949) has stated that in case the parent material is the dominant factor in soil formation, the most obvious way in which it influences is through similarity in colour and texture. This is applicable in soils of the present study as majority of them have brown or reddish brown colour and sandy loam/silty loam or loam texture. Dominant influence of parent material on the genesis of these soils is further supported by the similarity of clay minerals with predominance of illite followed by chlorite and vermiculite (Gupta and Verma, 1992) due to younger stage of parent material belonging to Middle Miocene, Pliocene and Lower Pleistocene period and weak chemical weathering to which these soils have been subjected. With time, however, such an effect may be obliterated.

Topography plays a significant role in controlling the rate of soil maturity. The exact role of this factor is difficult to explain, because of modification of original topography over a period of centuries. However, the soils developed under mild or gentle slopes (1 per cent or between 1–3 per cent) permit comparatively more percolation of water through solum and hence show the pedogenic development of colour and or structural B cambic (altered) horizon.

Although the influx of vegetation on the genesis of soils is not so perceptible yet the soils under forest vegetation have shown some profile development probably due to more content of organic carbon.

Classification

The soils have been classified according to U.S.D.A. System of Soil Taxonomy (Soil Survey Staff, 1975).

Ombrothermic data indicate that during monsoon season (July to September) the water supply is quite enough and the soils remain moist for 2 to 3 months in a year. The data further indicate that soils remain dry for about 7–8 months and moist for 4 to 5 months annually. All profiles except those under forests (12 to 15), therefore qualify for an ustic moisture regime. The mean annual soil temperature is more than 22°C and the difference between mean summer and mean winter temperature is more than 5°C, hence qualifies for hyperthermic temperature. The soil profiles under forests (12 to 15) can retain adequate moisture in the root zone for a longer period of time so as to qualify for udic moisture regime. The udic moisture regime implies that in most years the soil moisture control section is not dry in any part for as long as 90 days (cumulative).

The soil profiles 2 to 4 and 7 to 9 do not show much horizon differentiation owing to less maturity and hence have been classified into Entisols. Within the Entisols the soils of profiles 2–4 and 7–8 are classified into Orthents subsorder as organic carbon decreases regularly with depth and reaches a level of 0.2 per cent or less. Because of ustic moisture regime these soil profiles qualify for Ustorthents and Typic Ustorthents at great and subgroup level. The soils of profile 9 showing mottles with ehroma 2 qualifies for Aquic Ustorthents.

The soil profiles 1, 5 to 6, 10 to 12 and 13 to 17 show the development of altered structural B (cambic) horizon below ochric epidedon qualify for Ochrepts at sub-order level. The soil profiles 1, 5, 6, 10 to 11 are lacking of characteristics associated with wetness and volcanic matter and regular decrease in organic carbon with depth and reaches the level of 0.2 per cent or less and thus classified into Typic Ustochrepts. The soil profiles 16 and 17 have Ochric epipedon, calcareous throughout the depth and there is irregular decrease in organic carbon content but is higher than 0.2 per cent and slope of these profiles is less than 25 per cent. As such these profiles are classified as Fluventic Ustochrepts. Soils of udic moisture lack aquic or vertic chàracteristics and lithic contact and show regular decrease in organic carbon with depth, are classified as Eutrochrepts. Because of non calcareous nature the soil profiles 12 and 13 qualify for Dystric Eutrochrepts. On the other hand, the soil profiles 14 and 15 are calcareous and classified as Typic Eutrochrepts.

Land Use Consideration

Eutrochrepts and Ustochrepts occurring on nearly level to mild slopes (less than 1 per cent or between 1–2 per cent) have no erosion problems. Owing to the ustic/udic moisture regime, this area is suitable for growing of fruit trees like ber (*Zizyphus jujuba*), lime (*Citrus aurantifolia*) and guava (*Psidium guajava*) as well as *amla*

(*Emblica offcinales*). The cultivation of kinnow (*Citrus reticulata*) is also possible provided the provison of water is made for irrigating the kinnow orchards especially during the months of May to June and/or October to November. The Typic Ustorthents and Aquic Ustorthents occurring on the moderate slopes have high runoff and moderate erosion problem mainly during monsoon. In such areas, land use should aim at conservation of soil resources. These areas may be brought under grasses and fodder trees. Some of the tropical fruits like *amla* (*Emblica officinales*), guava (*Psidium guajava*) and *ber* (*Zizyphus mauritiana*) may also grow well in such areas.

References

Gupta, R.D. and Banerjee, S.K., 1991. *Problems and Management of Soil and Forest Resources of North West Himalayas*. Mahajan Book Centre, Gandhi Nagar, Jammu.

Gupta, R.D. and Verma, S.D., 1992. Characterisation and classification of some soils of Kandi Belt in Jammu Siwalik Hills. *J. Indian Soc. Soil. Sci.* 40: 809–815.

Gupta, R.D., 1992. Soil genesis in the Jammu and Kashmir Himalayas. In: *Himalayan Environment: Man and the Economic Activities*, (Ed.) J.L. Raina. Pointer Publishers, Jaipur, India.

Gupta, G.D., 1994. Characterisation of eroded versus non-eroded soils of Kandi Belt of Jammu District, J&K State. *M.Sc. Agric. Thesis (Soil Science)*, SKUAST, Srinagar.

Kaul, M.L., 1974. Pre-irrigation soil survey of proposed command area of Ravi Canal Project, Jammu, J&K. Soil Survey Organisation, Department of Agri., J&K.

Nazar, G.R., 1993. Characterisation and management of salt affected soils of Jammu district, J&K. *M.Sc. Agric. Thesis (Soil Science)*, SKUAST, Srinagar, J&K.

Robinson, G.W., 1949. *Soils: Their Origin, Constitution and Classification*. Murby and Co., 40, Museum Street, London, W.C.I.

Soil Survey Staff, 1975. *Soil Taxonomy: A Basic System of Soil Classification for Making and Interpreting Soil Surveys*. Agric. Handbook, U.S.D.A., p. 436.

Chapter 4

Impact of Degrading Habitat on Wildlife in Jammu and Kashmir

B.L. Kaul and Anil K. Verma

ABSTRACT

Kashmir Himalayas are divided into three constituent regions *viz.* the Sub Himalayan Jammu, the Himalayan Kashmir and the Trans-Himalayan Ladakh. These three regions can be likened to the three steps of a terrace. In recent past Kashmir Himalayas boasted of a well kept forest cover. Prior to 1947 these forests provided a habitat for a rich variety of wildlife. With the disappearance of forests resulting from mindless felling wildlife has been exposed to poaching. Consequently some of the most beautiful species of wildlife have become endangered.

Introduction

Pandit Jawaharlal Nehru's love of nature animated his very soul, reflecting a poet's dream. Mountains, valleys, trees, flowers and animals symbolized for him the multifarious and innumerable aspects of creation. Nehru's first visit to Kashmir after his marriage in 1916 cast a spell on his mind and soul which can be discerned in these words, " Like some supremely beautiful women whose beauty is almost impersonal and above human desire such was Kashmir in all its feminine beauty of river and valley and lake and graceful trees ". His love for the mighty Himalayas and wildlife is also well known, as he himself remarked that life would be dull and colourless without magnificent birds and animals.

Unfortunately, in a span of six decades after India gained freedom our hills and mountains wear a desolate look as a mute testimony to the virtual desecration of Nature's Charm and beauty. Much of the forest cover that used to be the pride of the lofty Himalayas and adorn them with rapturous charm, has disappeared, thanks to the unchecked activities of some unscrupulous people who through an indiscriminate felling of trees denuded the mountains and hills rendering them ecologically useless and counter productive. As a natural corollary to the ruthless deforestation, many species of wildlife in the Himalayas in general and in the Kashmir Himalayas in particular, are facing the threat of extinction.

The author would suggest that a serious study be made to co-relate the ecodegradation of a wide range of natural habitats to the problems connected with conservation of a variety of fascinating but endangered species of wildlife in Kashmir Himalayas. The study may prove beneficial in evolving steps towards the preservation of our precious wildlife heritage, besides being a humble homage to the memory of such a human and much splendoured personality as Pandit Jawaharlal Nehru.

Kashmir Himalayas

The three main constituent regions of the state of Jammu and Kashmir are the Sub Himalayan Jammu, the Himalayan Kashmir and the Trans-Himalayan Ladakh. These three regions can be likened to the three steps of a terrace, the lower step being the Jammu region, the middle one as the Kashmir valley and the upper one as Ladakh. In the Jammu region we have a range of hills known as *Siwaliks* and to the north of the *Siwaliks* are the middle mountains called as the *Lesser Himalayas* or the Pir Panjal range. The Kashmir valley is enclosed by the Pir Panjal range in the south and by the *Greater Himalayas* in the North. Ladakh is the land of mountain passes with the mountain ranges constituting the most striking features of the land. These ranges stretch from the south – east to the north – west and include the *Greater Himalayan Range* as well besides the *Zanskar Range*, the *Ladakh Range*, and the *Korakoram Range*. Many valleys nestle in the laps of these formidable mountain ranges.

Forests and Vegetal Cover

The type of vegetation found in Jammu and Kashmir is typical of western Himalayas, of which the state is a part. A largely mountainous terrain dominates the twin provinces of Jammu and Kashmir and rainfall is sufficient to favour growth of vegetation including trees. But in Ladakh region the cold and arid climate does not favour the growth of forests.

The outer slopes of Siwaliks in the Jammu region are covered with evergreen bushes and bear an open scrub forest. The slopes of the middle mountains are also clothed in temperate monsoon forests. Important species of trees found in the region are deodar, chir pine, blue pine, silver fir and oak. The mountain ranges surrounding the valley of Kashmir abound in deodar, blue pine, silver fir, birch trees and alpine brush wood of *Juniper* and *Rhododendron*. Besides these types of trees, willows and poplars also abound in the plains of the valley.

Before 1947, Jammu and Kashmir state boasted of a well managed forest wealth. However, after independence the forests which till then had been protected from the ravages of deforestation became victims of vandalism through an indiscriminate felling of trees. According to the figures available, the state has now more than 2 lakh hectares of degraded forests of which 1.10 lakh hectares are culturable and the rest constitutes a " dead loss ". This is without doubt a matter of great concern, not only in terms of economy but also from the ecological point of view.

With the fast disappearance of forests, wild animals too have fallen an easy prey to poachers and the trigger–happy hunters. The situation has already assumed alarming proportions, and some of the most beautiful and rare wild species like *Hangul, Snow leopard, Spotted deer, Musk deer* and *Himalayan Monal* are threatened with extinction in their respective habitats. Biologically it is a well known fact that a close relationship exists between an organism and its environment and any tampering with the habitat, automatically results in upsetting the natural life of the organism. Besides playing havoc with the natural habitats of wild animals deforestation in Jammu and Kashmir has also brought in its wake floods as a result of soil erosion on the mountain slopes. Construction of dams for generation of electricity alongside of major rivers for instance Chenab river in Jammu Province not only degrade the habitat of wildlife but also cause floods, soil erosion and landslides. Frequent land slides on the Srinagar–Jammu and Srinagar–Leh highways are also a direct consequence of ecodegradation of the Kashmir Himalayas, which has been going on unchecked for quite some time now.

Disappearing Wildlife

In the not so distant past, the state of Jammu and Kashmir boasted of a rich and varied wildlife. Unfortunately due to rapid depletion of the forest cover wildlife has been rendered vulnerable to reckless hunting and poaching. A stage has now come where some of the species of wild animals have come to the verge of extinction thereby finding inclusion in the red book of International Union of Conservation of Nature (IUCN). Unless immediate remedial steps are taken on a priority basis, we shall soon lose some of the finest wildlife specimens in our part of the Himalayas. Our state has the unique distinction of having the " Hangul " (*Cervus elephas hanglu*) found nowhere else in the world and it has already become an endangered species. Similarly, Snow leopard (*Panthera uncia*), Markhor (*Capra falconeri*), Musk deer (*Moschus moschiferous*) are other rare and beautiful animals facing danger of extinction.

Due to the varying type of terrain and climate, ranging from the plains of Jammu and alpine forests and shallow water bodies of Kashmir, to the lunar type mountains of Ladakh a rich bird fauna abounds in Kashmir Himalayas. These birds include the beautiful pheasants, partridges and water birds. Partridges and pheasants remain confined to dense forests ranging in altitude from 3000 to 15000 ft. above the sea level. Some of these birds like Western Tragopon (*Tragopon melanocephalus*) and cheer pleasant (*Catreus wallichi*) have been listed among endangered species in the red book of I. U. C. N.

What can be done?

Although fragmentary accounts are available on the geomorphology, vegetation, fauna and wildlife in Kashmir Himalayas, yet so far no attempt has been made to have a realistic and comprehensive overview of the situation as it exists today. The available data are scattered in different journals and periodicals and there is no single publication to which a layman or even an expert can refer. Further, there has been no serious attempt to co-relate the ecodegradation with the existence of wildlife in Kashmir Himalayas. As such the authors suggest that the following aspects of the problem of conservation of wildlife be studied in detail:

1. Assessing the extent of ecodegradation of natural habitats pertaining to the wildlife of Kashmir Himalayas.
2. Collection of information related to present status and distribution of endangered species of mammals and birds.
3. Study of Predator–prey distribution in natural ecosystems and the role played by man in disturbing this relationship.
4. Collection of data regarding viable breeding populations of endangered species.
5. An assessment of the role played by man and natural agents in ecodegradation of the habitats of wild animals.
6. Cataloguing the rich biogeographic diversity of wildlife in Kashmir Himalayas, which is a combination of Oriental and Palaearctic fauna.
7. Taking remedial measures for restoration of habitats and conservation of endangered species of mammals, birds, herpetofauna and butterflies.

References

Atul and Punam, 1989. Potentials of agroforestry in Himachal Himalaya. In: *Environmental Hazards in Himalayas*, (Ed.) S.K. Chadha, Pointer Publishers, Jaipur, pp. 37–56.

Chadha, S.K., 1989. *Environmental Hazards in Himalayas*. Pointer Publishers, Jaipur pp. 1–13.

Dar, Moti, 1987. Ladakh. In: *Wildlife of J&K*. Sanctuary Publications.

Ganhar, J.N., 1979. *The Wildlife of Ladakh*. Harmukh Publications, Srinagar.

Gergan, S.S., 1962. Big Game of Jammu and Kashmir. A Government of J&K Publication.

Kaul, B.L., 1989. *Ecology and Wildlife of Kashmir Himalaya*, (Ed.) S.K. Chadha. Pointer Publishers, Jaipur.

Kaul, B.L., 1989. Ecology and wildlife of Ladakh. In: *Environmental Hazards in Himalayas*, (Ed.) S.K. Chadha, Pointer Publishers, Jaipur, pp. 66–74.

Kaul, B.L., 1994. Call of the Wild. *The Hindu*, Chennai, May, 1.

Chapter 5

Impact of Tourism and Development on Ecology and Biodiversity of Patnitop Hill Resort

A.K. Parimoo

ABSTRACT

Tourism has emerged in recent times as a major pleasure and economic activity in India and the hill resorts in the Himalayan States are a preferred destination both for the foreign and domestic tourists. However, tourism has inevitably affected the ecology and biodiversity of these resorts. While growth of tourism for economic progress is essential for the welfare of the local people, conservation of the ecosystems has to be taken care of at all costs. In recent times pilgrims to Mata Vaishno Devi Shrine have also started visiting the neighbouring hill resorts of Patnitop, Kud, Sanasar and Batote. This has adversely affected ecology of these hills in the middle Himalayas. In order to conserve the ecology, flora and fauna and to help in development of the area Patnitop Development Authority has come into being. The study concerns the impact of tourism and development activities and measures required to be undertaken for conservation of the ecology and rich biodiversity of the area.

Introduction

The beautiful Patnitop area broadly covers 265 sq.km. spreading over parts of two districts *viz.*, Udhampur and Doda of Jammu

Division. A comprehensive and holistic development, especially in order to enhance not only the growth of tourism but also other social and physical facilities including conservation of flora and fauna within the area with a sensitive approach has been going on under the Patnitop Development Authority. This area extends from the tehsil headquarter of Chineni from the south to the administrative town of Batote to its north. Development is also envisaged at the major settlements lying along north of river Tawi upto village Dudu and further uphill to Basantgarh, the ancient abode of the king of Kashmir.

The area includes places that have gained a considerable level of development *viz.*, the hill resort of Patnitop and nearby places like Sanasar, Kud and Batote. Development has primarily taken place owing to their location alongside Jammu Srinagar Highway (NH 1A). On the other hand, areas away from this highway have not attracted much development, most of which are yet to be thrown open to tourists. Basantgarh is to be made accessible by vehicular roads from Dudu side by putting a bridge across the river Tawi.

The entire area is predominantly rural in character except Batote, a small town and a traditionally important halt on Jammu Srinagar National High way. Chineni, another small town is situated at a strategic location on the foothils side of the NH from Jammu and is just outside the Patnitop Circuit Area. The entire area under Patnitop Development Authority has been divided into three broad zones (or subcircuits) based on physical proximity, accessibility and prevalent trends of tourism in order to achieve an effective and phased development.

While development of the area is essential for the growth of tourism and economic progress of the local people, yet the ecology and environment are to be conserved at all costs. The beauty of the area is mainly due to its flora and fauna which need to be conserved to retain its charm.

Climate

There is a wide variation in temperature and climate. The temperature varies between 42°C and –5°C. It has been observed that there is a fall of 1°C atmospheric temperature in the hills for every 270 m rise in altitude upto 1500 m beyond which the fall is more sharp. Since there is an altitudinal variation of approximately,

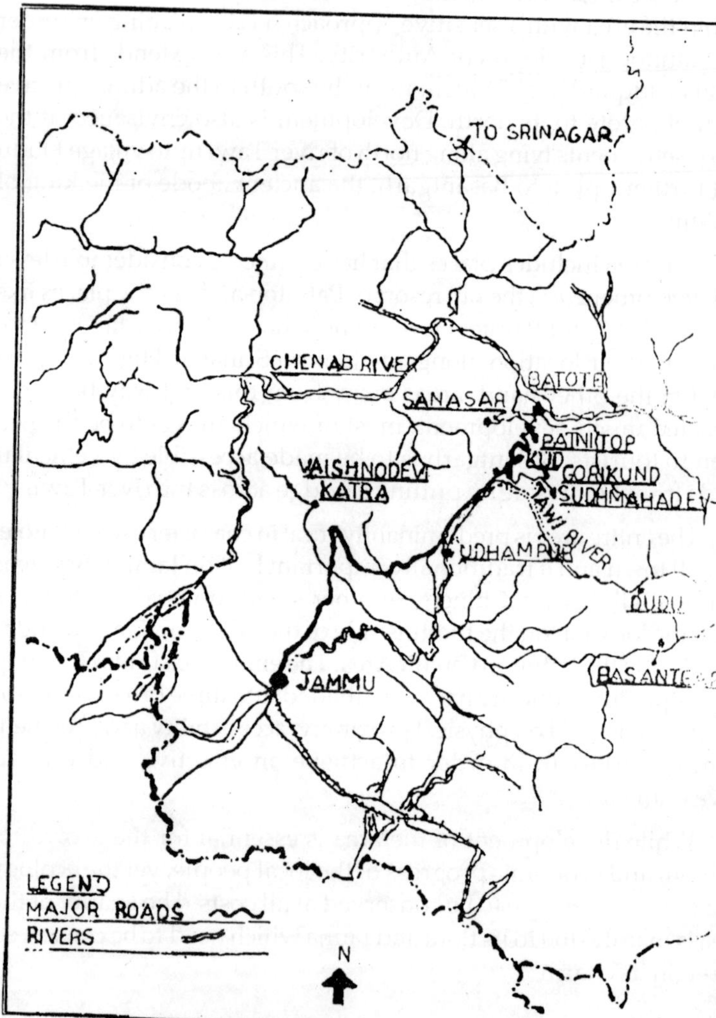

Figure 5.1: Map Showing Sudhamahadev, Patnitop, Kud, Batote, Lati and Dudu

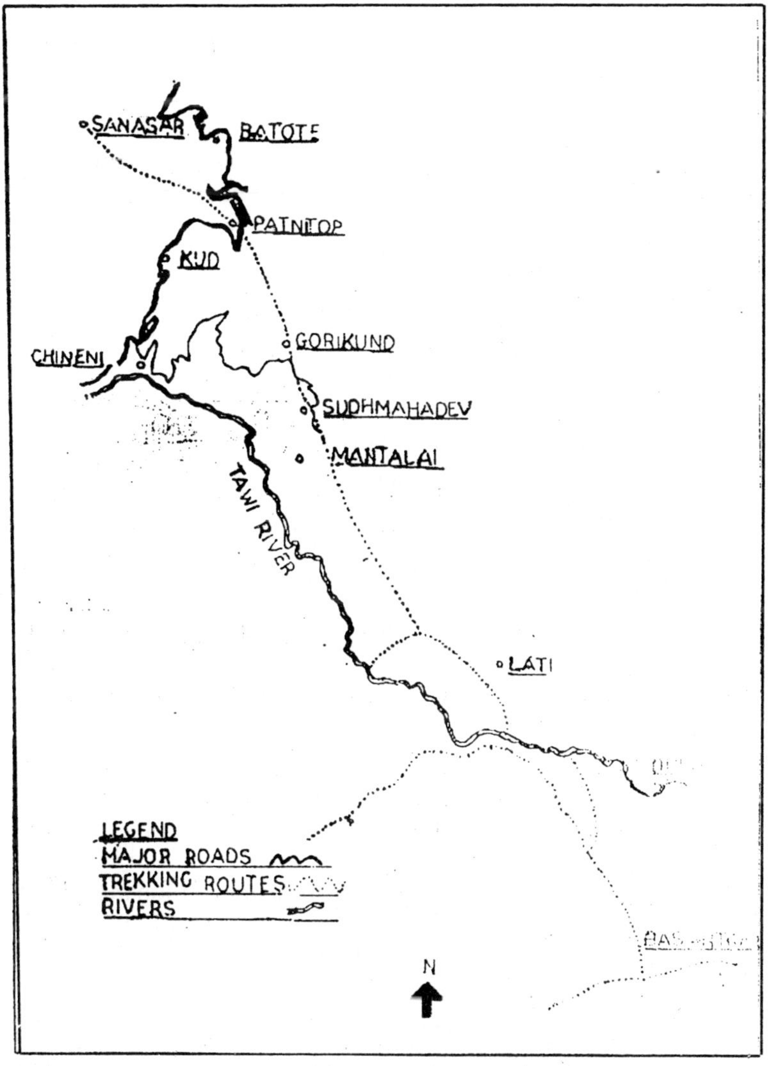

Figure 5.2: Site Plan of Patnitop and Lati

3,300 m a minimum difference of 14°C or more can be expected at any given time. December, January and February are the coldest months. July to September is the rainy season.

Snow fall is a regular feature at higher elevations during winters. Summer is reasonably hot in the low lying places reaching a peak in the month of May and June.

The natural agents such as wind, snow avalanches, land slides, drought and lightening render considerable damage to the forests at places particularly to the young flora and the trees.

The circuit gets the benefit of both the winter and monsoon precipitation from December to March and July to September respectively. A decadal rainfall information for the year 1974 to 1984 has been assessed and the major observations are:

1. The circuit receives major portion of the precipitation during the winter months of January to March, both in the form of snow in higher regions and rains in low lying areas. The precipitation is mostly in the form of rains during summer months of July to September. The intensity of precipitation during the winter months is higher than that of summer.
2. April to June and September to December are comparatively drier months in the year. This period as such warrants constant and regular vigil against threat of occurrence of the forest fire.
3. The distribution of precipitation over the entire tract has remained almost the same over the past 10 years. However, the intensity of precipitation increases from South to North.
4. Cloud bursts which were rare and uncommon in the past, have now occurred at times and a few pockets suffered very badly during the past decade. The frequency of such bursts is on the rise.
5. Frost and mist are uncommon within the region as compared to the Kashmir valley, Sunny days are warm and occur frequently. Radiation frost is common in the region.
6. Droughts are of rare occurrence in the region.

   A sizeable chunk of the cultivable lands are situated on the mountain slopes which are sensitive to erosion.

Moreover, the rains are sometimes so heavy that the entire crop on the slopes is washed away. In such circumstances the economic life of the villagers comes under heavy strain.

The Forest Cover

The forests of this division are predominantly coniferous. Altitude-wise we find the species Deodar, Kail, Fir and Chir in that order.

Deodar

Deodar are evergreen trees with fairly well developed dark green silvery canopy. It is predominantly seen at Kud and extends eastwards till Dudu and beyond. It forms a high density over the northern slopes. It covers about 6760.04 Hectares of the area. It is found in association with other species like Walnut, Poplar, Horsechestnut etc. as under growth. The most valuable and commercially important species of Deodars are found in areas having 1700–2500 m elevation. Deodar is found on more or less easy slopes and well drained loamy soils. Deodar accounts for 40 per cent of the total forest area.

Kail

Kail is the leading conifer, in its establishment on the fresh alluvial soils. In the Dudu valley which starts from Chineni and extends in far east. Kail occupies the left half and is found on an elevation of upto 2740 m (9000ft) on warmer areas growing side by side with Fir and Oak. It is an evergreen tree with conical/spherical/ crown which appears pyramidical in shape from a distance. In the study area it is confined to the upper catchment of river Tawi in Dudu and Udhampur ranges, where it is seen as a compact and dense patch. It accounts for 0.1 per cent of the total forest area.

Fir

Fir zone lies between 2135–3700 m and fir forests cover most of the catchment of the Tawi River and its side tributaries such at Lati nallah and constitutes more or less compact, well stocked fine stands. The younger generation of fir either constitutes a smaller percentage of the total or is totally absent.

Chir

The Chir grows between elevations of 520 m to 1070 m. It is mainly grown in the low lying areas of Udhampur and Ramnagar areas. Good quality Chir can be found in Batote Range and around the region south of Chineni starting from Kud to Dudu.

Pine

Pine species are found at altitudes of 1200 m–1800 m confined to the inner valley of Chenab and Tawi rivers and their tributaries. Most of this category is dense unless thinned out by fire or mismanagement and normally such dense groves have little undergrowth. The tree height often attains 20 m to 30 m. Pine forms one of the important forest types in the study area and occurs over Udhampur and Ramnagar ranges. Good quality of Pine occurs in and around Chineni. Pine species cover about 4854.92 hectares of the total forest area.

Oak

Oak groves are seen in pockets near Devi Ka Poddar forest east of Batote, at an altitude of 2400 m North-East and East of Sudh Mahadev and Gorikud at an altitude of 2300 m, North of Tawi in varying densities. It accounts for 2 per cent of the total forest area in the circuit. Oak is in demand by the locals for the supply of fodder, fuelwood and for manufacture of agricultural implements. Demand for fuel and fodder and cultivation on gentle slopes led to extensive felling of the species. This is very much evident from the fact that now oak accounts for less than 10 per cent of the crown density.

Banj

This species is of moderate size and is noticed along the valleys and at the foothills where cultivation is practised. Well development of Banj forests is observed in south of Chineni on the way to Dudu and Basantgarh. In Mantalai these species are seen in conspicuous patches though many of these are bare due to severe logging and cutting for fodder. These species cover an 11 per cent of the total forest area.

Scrub Forests

The scrub forests are confined to low lying plains of Udhampur and Ramnagar ranges and are found as broad leaved associates of upper and lower limits of Chir forests. These forests have been subjected to heavy grazing and excessive unregulated felling in the past. This has resulted in the formation of permanent blanks and excessive soil erosion. These forests meet the fuel and fodder needs of the locals and if managed properly on modern silvicultural lines can provide raw material to meet the requirements of wood, sports goods and furniture industry.

Patches of scrub forests are seen to the east of Dudu at an elevation of 2300 m and are confined to small pockets. Scrub forests are noticed on the south western slopes where the elevation of southern side is higher and has sheltered the area in the eastern position. Some patches are noticed even close to the snow covered area.

Mixed Forests

Mixed Forests account for 14 per cent of the total forest areas. These areas are mostly disturbed by human intervention and are seen south west of Udhampur on the way to Dudu and Basantgarh.

Medicinal Plants

The Patnitop area abounds in a variety of herbs, roots, rhizomes and bulbs which have medicinal value. These need to be conserved for proper exploitation and benefit of mankind.

Forest Plantation

Extensive plantation areas are seen west of Patnitop and south of Chineni and is mostly composed of pine. It accounts for 0.6 per cent of the total forest area.

Agriculture within Notified Forests Boundary

Areas under agriculture within the notified forest areas are scattered throughout the study area in segregated patches and account for 2 per cent of the total forest area.

Existing Problems

The problems that are encountered in these forest areas are as follows:

1. *Fire: both natural and artificial*: It has been the major problem in managing forests especially that of Chir, Deodar and Kail forests. Frequent fires accelerate the soil erosion and lower the soil quality.

2. *Expansion of man's habitat and animal grazing*: Uncontrolled and unrestricted grazing by all kinds of animals, both at high and low level forests hinder the natural regeneration. Excessive grazing causes hardening of soil and hampers germination of seeds thus inhibiting natural regeneration.
3. *Lopping*: Lopping for fodder and fuel wood is very common in such forests which occurs in the vicinity of human habitation. Reckless lopping of oak forests in the past have deteriorated them into oak scrubs.
4. *Torch wood extractions*: It is done by villagers to extract resin laden small sticks from standing chir trees which are used for lighting. As a result these trees are damaged and get weak and ultimately break from the stump level during storms.
5. *Irregularities in resin extraction*: During the extraction of resin, the prescribed norms of depth, length and breadth of blaze are invariably exceeded. Such excesses with respect to size of a channel blaze weakens the younger crop and make them susceptible to wind breaks especially along the spurs and ridges.
6. *Encroachments*: Encroaching upon the forest lands adjoining their cultivation by merciless felling or girdling of the tree growth is a common practice among villagers.
7. *Illicit felling*: Despite several concessions the villagers cause considerable damage to forests by resorting to illicit felling of trees and smuggling of timber which coupled with the population explosion, rise in the living standard of villagers and rapid urbanisation pose a threat to the forests. In many ways, conservation of the forests and strict protection of the species with their diversities and mass awareness are priority tasks.
8. *Girdling*: These damages are noticeable around Chaks.
9. *Wild animals*: Black bears cause girdling of deodar, kail and spruce poles. Porcupines uproot and chew the roots and seedlings of chir. However this damage is comparatively negligible.
10. *Insects, fungi and parasites*: Insect attack is not so common in this region. Certain scolytidae insects attack dead and

diseased trees of deodar; the cicadids damage the kail forests.

Environmental Threat Caused by Market Demand on Forest Produce

Today, the region is on the threshold of severe environmental degeneration as a result of the large scale economic pressure coupled with an indifferent attitude of the responsible authorities. The degradation of the environment is mainly caused due to the large scale exploitation of the natural resources occurring in the mountain tract with the active connivance of forest officials.

Timber and resin are the two main forest products of the area. The timber is extracted from all over the Jammu Province. The crude resin extracted from the low lying chir forests, is the other most important forests produce, as it is a source of substantial revenue to the forest department. However, as has already been mentioned, the prescribed norms and techniques for extraction are rarely followed which results in weakening of the crop and further degeneration.

The Wildlife: A variety of fauna is found in this division because of the varied climatic conditions and altitudinal variation prevailing in the tract. The faunal species (mammals and birds) present in the area are:

I: Mammals

1. The Leopard or Panther (*Panthera pardus*): It is commonly known as "Chita" or "Chitra" by the locals. It is mainly confined to a small portion of sub-alpine and alpine belts mostly close to the villages. Many a time it kills domestic animals, with the result that locals hunt this animal largely for protecting their animals. Its number is on a sharp decline. It has been declared as Special Game as per J&K Wildlife Protection Act 1978.
2. The Snow Leopard (*Panthera uncia*): A few individuals of this animal are surviving at high altitudes. It generally inhabits the snow mountains and has been included in Schedule I of J&K Wildlife Protection Act 1978 and its killing is totally banned.

3. The Leopard Cat (*Felis bengalensis*): This forest cat is now almost extinct. It is nocturnal in habit, rarely seen during the day time.
4. The Jungle Cat (*Felis chaus*): A small number of this species is generally found in lower scrub forests of this division.
5. The small Indian Mongoose (*Herpests auropunctatus*): It is mostly found in low lying area of this division in and around cultivated fields.
6. The Jackal (*Canis aureus*): It is mostly found in low land around towns, villages and cultivated areas sheltering in holes in the ground dense grass and scrub. It is nocturnal.
7. The Red Fox (*Vulpes vulpes*): It is usually found in the upper sub-temperate and temperate zones of the area.
8. The Indian Fox (*Vulpes bengalensis*): It lives in burrows dug by itself in open ground or in scrub.
9. Indian Wild Dog (*Canis alpinus*): This animal is occasionally found in the forests of this division.
10. The Himalayan Black Bear (*Selneractos thibetanus*): It is now found in the interior steep forest area of Lander and Marmat ranges.
11. The Common Otter (*Lutra lutra*): This animal is occasionally found in the streams and springs throughout the division.
12. Barking Deer (*Muntiacus montjak*): It is found in very few numbers in low lying Chir bearing area of Batote.
13. The Indian Wild Boar (*Sus sciofa*): The wild bear once existed in large number but is now reduced to a few ones found mostly in the subtropical belt of Batote and Lander ranging upto an altitude of 2,500 m.
14. The common Langur (*Presbytes entella*): Known as langur it is commonly seen in the forests near Udhampur.
15. The Rhesus Macaque (*Macacca mulatta*): Locally known as bandar, it is found almost all over in the low lying area upto 2500 m altitude.
16. The Red Flying Squirrel (*Petauista phillippens altreventa*)
17. The Indian Field Mouse (*Mus* sp.)
18. The Indian Porcupine (*Hystrix indica*)
19. Brown Bear (*Ursus arctos*)

20. The Grey Himalayan goat (*Nemorchaedus goral goral*)
21. Musk deer (*Moscus moschifera*)

II. Birds

1. Monal Pheasant or Impejan Pheasant (*Lophophorus impejanus*). Found in high level Fir and sub-alpine zone of Marmat range in summer and descends down in winter. It is occasionally seen in the Batote ranges.
2. The Koklas Pheasant (*Ceiottis mauolopus*): Mainly found around Batote Range.
3. The Grey Partiridge
4. Jungle crow
5. Roller or Blue Jay
6. Koel
7. The Himalayan tree pie
8. Indian Myna
9. Small yellow wood pecker
10. Golden backed wood pecker
11. Roseringed parakeet
12. Indian robin
13. Tailor bird
14. Paradise flycatcher
15. Aquatic Birds

A few species of migratory ducks are encountered seasonally along the river Chenab and wetlands.

III. Reptiles

Variety of snakes, both poisonous and non poisonous are found all over the division. Among the poisonous snakes Cobra, Viper, Krait and Pitviper are more common. Agamid and Varanid lizards are common in subtropical belt.

IV. Fishes

Local variety of fishes including Mahseer, *Schizothorichthyes* and exotic Brown trout are of common occurrence in the river Chenab its tributaries and Tawi river.

Problems Faced

Various reasons, especially that of ever increasing pressure of human population and tourism are responsible for a sharp decline of wild animal population in this region. The worst sufferers are mainly game animals and birds. Excessive interference by the graziers and their large herds of grazing animals in the alpine and sub-alpine zones have led to large scale destruction of certain species of wild animals like Wild goat, Musk deer, Brown bear, etc. Deforestation has also largely been responsible for disturbing the habitat of the wildlife and sharp reduction in their numbers. Road building and ever increasing trekking have also threatened wildlife.

Injuries to which Fauna is Liable

The fauna of the tract is liable to certain injuries such as:

(*i*) Injuries by Man

Hunting (both legal as well as illegal) of wild animals and birds has always been game for the man. Wild animals and birds are being killed for their valuable skins, horns, flesh etc. At times this also has created an ecological imbalance in the biological pyramid by depletion of certain forms of wildlife. Large-scale destruction of forests by man due to encroachment, excessive felling and frequent fires is also responsible for destroying the habitat of the wildlife.

(*ii*) Injuries by Epidemics

Contagious diseases spread among the wild animals mainly through the domesticated animals grazing inside the forest thereby depleting considerable faunal population.

(*iii*) Injuries by Fire

Forest fires are responsible for large scale destruction of birds and microfauna. At times wild animals also get trapped and killed in the forest fires.

(*iv*) Injuries by Atmospheric Influences

Atmospheric influences do affect the wild animals and the birds specially the young ones. The birds do suffer from heavy snow fall, rains, storms and drought as their young ones and eggs are destroyed by such influences. Under normal circumstances many forms of wildlife live on other types, a natural process which controls the balance of the wildlife. With the result all forms of the wildlife continue to co-exist without disturbing the ecological balance.

Environmentally Degraded Area

Patnitop-Sanasar Area

Land along the Patnitop-Sanasar road and the adjoining areas are found to be one of the most environmentally degraded areas in the entire Patnitop circuit. Large portions of these lands are bare and the upper section of the Sanasar area is seriously degraded. Cutting of forests on lower and steep slopes, overgrazing on steep slopes and also alignment of the Patnitop Sanasar road itself have contributed to its present deteriorated situation.

A large area is under steep slopes and appropriate preventive measures have been taken to check the soil erosion. Land slides are now common in the area along the upper portion of Sansar which is almost bare leading to heavy soil erosion during the rainy season. The area, once covered with thick forests of deodar, fir and kail, is now facing serious environmental problems. Deodar trees specially those near the settlement, are dying and kail trees above 4000 feet are becoming extinct.

Fir trees, are also diminishing in number due to degradation and grazing as a result of biotic and unplanned tourist activities. Due to degradation of the area the seeds that fall during the summer season are washed away by rains. As these areas are not properly fenced and demarcated, people walk on them thus the growth of vegetation and trees is reduced.

Steps to Prevent Degradation

Some steps are underway for soil conservation in the area. Curative work which involves dressing of the rocks and slopes is one such step that has been taken along the Sanasar lake and along many other natural springs of the area.

Vibernum nurosum which is a bushy plant and a good binder has been planted in several places. It enables easy construction of roads and other built up areas which lead to soil and land erosion. So, preventive measures should be taken to preserve it.

Since it grows fruit during the summer months of May and June fencing becomes necessary to protect the area. By fencing or artificial plantation further promotion for the growth of seeds and hence the trees can be effected.

Sanasar Lake

The high altitude wetland situated 19 km North West of Patnitop has undergone severe degradation. The area of the wetland is 2.95 ha with an average length and width of 825 feet and 385 feet respectively. Surrounded by large tracts of deodars and kail forest, a considerable area has been largely degraded due to agricultural activities and other biotic interferences. The lake once sustained a variety of fish fauna, water fowl, aquatic birds and other organisms and this erstwhile clear water body has been rendered into a degenerated marsh. The lake is poorly maintained and it seriously reduces the potential of a nodal spot for the tourists as well as a component of good natural environment. Presently, the surface of the lake is highly turbid accompanied by extensive algal generation.

Already an eco-development programme for Sanasar has been drawn up by the Department of Environment and Remote Sensing with the following objectives:

1. Reconversion of Sanasar marsh into a fresh water natural lake.
2. Maintain physical, chemical and biological characteristics of the lake.
3. Reverse the process that has led to the degeneration of the lake and catchment.
4. Conservation of ecological assets of the area compatible with the ecological requirements for tourism promotion and comfort.

The details of the above programme is as follows:

(*a*) Reconversion of Sanasar lake into a natural fresh water Lake.

(*b*) Selective manual weeding: Large macrophytic vegetation is present in large quantities in marsh. Consisting of the dominant among them are *Ceratophylum* spp., *Polygonum* spp., *Myriophylum spicatum, Potamogeton* spp. and a number of other forms. In selective deweeding all considerable weeds will be removed except few useful ones for seeding.

This requires deweeding of 0.0295 sq km of marsh with vegetation depth of 1.9 ft.

(*c*) Desilting: Which aims to clear the lake of excess silt and achieve an average depth of 6 ft. of clear water. It requires desilting of 3 ft deep silt sedimentation.

(*d*) Maintenance of water level, circulation and flushing. Inflow and outflow channels need to be provided to maintain regular circulation of water and affect flushing of lake at regular intervals.

Inflow channel is to be provided with origin in the catchment area and enter the lake at northern extremity. It will carry water from catchment area into lake after passing over check dams and other silt retention structures. The outflow channel will be provided at north eastern extremity of the lake which will drain out excess water outside the catchment area. Flush should be provided at the mouth of the channel to flush the lake.

Measures in catchment area to prevent sedimentation of the Lake

The catchment area of the lake has degraded mainly due to biotic interference and extensive agricultural practices. It has been found that a major nallah taking waste water off the area directly drains into the lake. Cultivation of slopes without proper terracing as well as application of fertilizers and pesticides has resulted in eutrophication and sedimentation of lake.

Therefore, it is intended:

(*a*) To take up soil and water conservation measures in the catchment area.

(*b*) To ensure effective sedimentation of wastes in the nallah by providing check dams and other slowing measures.

Role of Grazing in Degradation of Natural Habitat

Livestock is owned for the livelihood, transport and other economic considerations. But for its maintenance it is the forests which are subjected to uncontrolled, haphazard, unregulated and disproportionate type of grazing. During summer, the sheep, goats and cattle belonging to bakerwals and locals migrate to behaks and alpine pastures while with the advent of winter they come down to plains of Udhampur and Ramnagar. The incidence of grazing is very important factor which has got direct impact on the vegetation. It affects the management if it is uncontrolled and unregulated.

The unrestricted, uncontrolled and unscientific grazing in the forests is largely seen to be responsible for degradation of fir and chir forests. Over-grazing is the main cause for failure of regeneration and loss of flora and fauna.

Strategy for Conserving Biodiversity

In the wake of growing tourism and development preventive measures are required for conserving ecology of the area including embankments and afforestation works to be taken up in the area prone to landslides to check their activity and soil erosion. The landslides can be controlled by protecting the slope. For protection from slides caused by scouring action of torrents, a stout retaining wall needs to be constructed at the point where the slope meets the channel. The water from the catchment area that passes through a landslide zone should be diverted so that the soaking of the bed rocks and the wetting of the clay is prevented. This can be further stabilised by planting suitable species and by wetting.

Construction of small embankments along the main stream along suitable locations will reduce soil erosion and siltation of rivers, and increase the ground water potential. Water spread in broad valleys and flat areas will add to the aesthetic beauty of the area.

Wasteland areas should be taken for development by horticulture through mixed plantations. It is noticed that some wasteland areas are already being developed at Chineni.

For the purpose of treatment the areas allotted can be classified into the following categories:

(*a*) *Areas bearing Oak forests and other broad leaved species in pure or in mixture with Chir*: The most important treatment required at present is strict, complete protection against wanton reckless lopping and felling by local population. The *Oak* forests are very vital for soil and water conservation and are responsible for maintaining regular flow of water in the springs and streams in the lower areas.

Many of the prominent Oak areas are bearing the brunt of meeting the ever increasing demand of firewood, and are getting degraded. Along with effective protection the artificial regenerative measures should be adopted on an extensive scale to rehabilitate these areas.

(*h*) *The areas covered with scrub forest and devoid of any tree cover*: The extent of the area under this category is 13.6 per cent of the total area of the working circle. This requires treatment on top priority as any further delay will make the condition still worse. Location of these blank/scrub areas varies from sub-tropical to temperate zone. In most areas soil has degraded to such an extent that it can no longer afford any vegetation cover. Hence soil status should be upgraded by undertaking soil conservation works in these areas. The soil conservation measures will increase the water retention capacity increasing the soil depth.

Realisation for Yield

As far as possible only dry trees should be marked to cater to the needs of locals and only in very rare case the felling of green trees should be allowed.

Aforestation Programme

This programme has 3 limitations *viz.*, average size of funds, planting stock and period for which an area can be closed to grazing at a time. The social forestry programme started recently in the state has an inseparable component of rehabilitation of degraded forests.

General Prescription

In the case of Chir forests where accidental fires are the major threat to the survival of regeneration. The collection and controlled burning of slush and dry pine needles, construction of firelines etc. are some of the fire protection measures, recommended for the areas.

Resin Industry

Resin extraction is mainly done from Chir forests of Udhampur and Ramnagar ranges. This is sold locally and in other part of Jammu province. The system is most injurious to the future growth of the tree.

Conclusion

The thrust of the strategy for the development of the various identified settlements should be in harmony with maintenance of ecological balance, promotion of quality of life of the hill people and conservation of biodiversity. The major area of development being socio-economic growth, a judicious utilization of land and natural resources needs to be integrated with development of infrastructure and promotion of ecology. Tourism as an industry should be encouraged with due regard to minimum exploitation of the scarce local resources including the flora and fauna.

All hilly regions in Himalayas are important both from ecological and economic point of view. Hills are national assets and constitute the important mountain ecosystems. However, it is unfortunate that the flourishing status of enormous hills resources have been considerably degenerated because of (*i*) poor socio-economic conditions of the local people, (*ii*) population stress and pollution problem, (*iii*) random deforestation, (*iv*) over exploitation/ poaching of precious wild plants and animal resources, (*v*) unscientific approach in hill development projects, (*vi*) unscientific agro-farming, agricultural management practices and (*vii*) improper tourism management. Hence proper care should be taken while developing the settlements in the Patnitop Circuit in order to prevent environmental degradation and loss of biodiversity of the area. This can be achieved by a careful and thoughtful planning for future development.

Chapter 6

Endangered Wildlife of Trans-Himalayan Ladakh

B.L. Kaul

ABSTRACT

The most striking feature of Ladakh Himalayas are breathtakingly beautiful mountain ranges, water bodies and varied and rich wildlife. Ladakh has very low population density *i.e.*, 3 persons/sq km in Leh District and 8 persons/sq. km in Kargil District. The fauna of Ladakh is a mixture of Palaearctic and Oriental regions. The most important and notable elements are mammals – mostly ungulates, rodents and carnivores–and birds including the famous "tung tung" or black-necked crane.

Introduction

Ladakh (La- pass; Dakh–land) is the land of the passes. Comprising of the districts of Leh and Kargil, it is the largest in area of the three main regions of Jammu and Kashmir state. Its most striking feature are the mountain ranges which stretch from the south-east to the north-west. Ladakh lies between 32-36°N and 76-79°E. Although the most part of Ladakh is mountainous there are many beautiful valleys lying in the lap of mountain ranges which include the greater Himalayan range, the Zanskar range, the Ladakh range and the Korakoram range. Like all high altitude mountain ranges of the world Ladakh is sparsely populated being inhabited by about 2.36 lakh people. Ladakh is divided in to two districts namely Leh district with population of 1.17 lakh, Kargil district with a population of 1.19 lakh. Population density in Leh district is 3 persons / sq. km and in Kargil district 8 persons/sq. km. Life is difficult at a high altitude of 2900–5900m but both man and beast brave it.The rich variety of wildlife in Ladakh bears out this fact.

Landscape of Ladakh includes snow-clad mountains, treeless arctic deserts and a few narrow fertile valleys, some of them breathtakingly beautiful. With the advent of summer a miraculous change occurs when life – sustaining vegetation lends colour to an otherwise bleak landscape. Ladakh by and large is a large barren but beautiful region of intense sunlight and clear unpolluted air. The temperature in summer rises to 35° to 37°C but drops to –30° to 40°C in winter. Dras in Kargil district records the lowest winter temperatures.There is very low rainfall (92mm) because it is situated in the rain-shadow of the Himalayas.

Most of the water available in Ladakh is in the form of frozen snow or ice. The great river Indus flows from its origin in Mount Kailash over the Tibetan plateau through narrow gorges and finally enters the Korakoram range. In Tibet the Indus is called Singhchu (Singh–Lion; Chu – mouth) as it is believed there that the river originates from the mouth of a lion. Singhchu has been distorted to sindhu, Indus and finally India. The word "Hindu" itself is derived from Sindhu. The other rivers of Ladakh region include Nubru, Shyok, Dras, Suru and Zanskar. All these originate from Northern glaciers and finally join the Indus which then flows into Pakistan. There are many brackish water lakes in Ladakh. These include the famous Pangong tso (tso–lake), Morari tso, Kar tso, Rul tso, Henle tso and Mitpal tso.

Ladakh is strategically as well as from wildlife point of view a very important part of the country situated as it is between Central Asia, China, Afghanistan and Pakistan. The wildlife in these parts of Asia is quite similar because of their proximity.

Rich Wildlife

Despite the extremely harsh environment and sparse vegetation wildlife in Ladakh thrived before independence. This was mainly due to Buddhist philosophy of non-violence. After independence Ladakh became more approachable. Construction of roads for defence purposes and development of the region provided greater

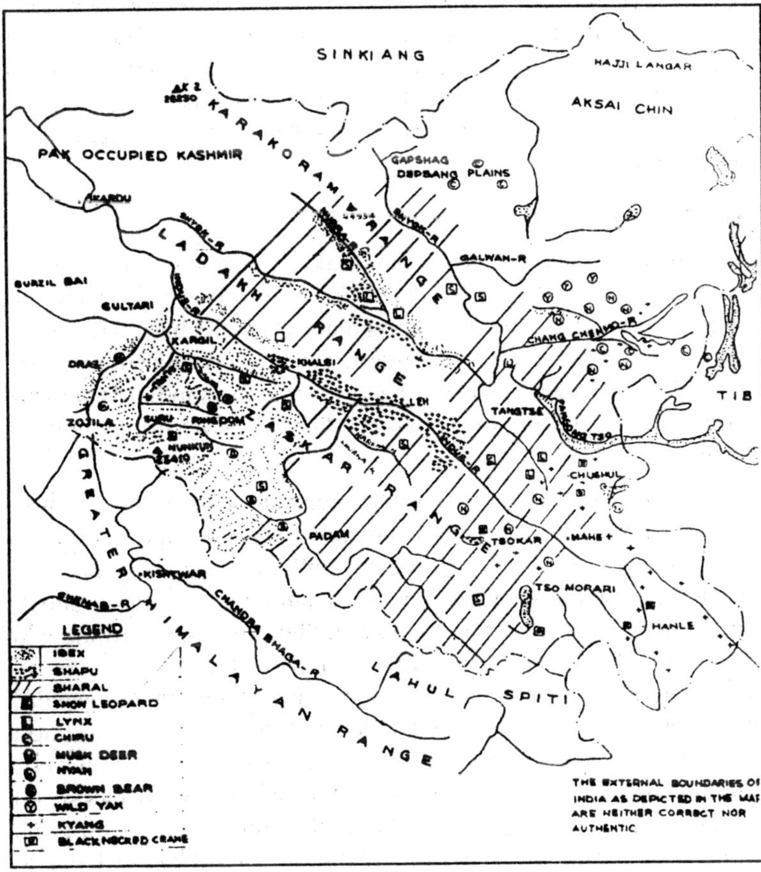

Figure 6.1: Ladakh Distribution of Mammals

opportunities for poaching and wanton killing of wild animals. The defence personnel also had a share in the killings because of their occasional indulgence in 'Shikar' a legacy of the "Raj" days. In short the opening up of Ladakh to rest of the world exposed its wildlife. Mercifully realization has dawned upon the government and other agencies of the danger of extinction of the varied beautiful animals in Ladakh and steps for their conservation have been initiated. The army personnel are also helping the civilian authorities in this noble task.

The fauna of Ladakh is a mixture of Palaearctic and Oriental regions. It comprises mostly of mammals and birds. The region is poor in reptiles and amphibians. The mammalian population consists mostly of ungulates, rodents and cornivores. The avifauna includes birds of prey, finches, plovers, larks, geese, grebes, pigeons, snowcocks, partridges, gulls, ducks and the famous "tung tung" or black-necked crane. A short survey of the mammals and birds is given below.

Mammals

The most noteable animal of Ladakh, Tibet and adjacent parts of central Asia is the wild yak (*Bos grunniens*). It is a large animal weighing about a tonne. It has long black hair which is tinged with grey at the muzzle. The horns are long and curved which may grow upto 75 cm in length. The wild Yak which moves in herds spends summers at great heights and descends down to low altitudes in the winter.

Another pride of Ladakh is the nyan also called Tibetan sheep (*Ovis ammon*). A large male may weigh as much as a quintal. These sheep possess curved horns about 150 cm. long. They move in herds of a dozen or so and spend most of their time at heights above 4500 meters. Dar (1987) has estimated their number to around 200 in eastern Ladakh. A close relative of nyan is Shapo (*Ovis orientalis*). It is smaller than the nyan, a large ram being around 80 Kg in weight. It prefers heights between 3000-4000 meters. The number of shapo in Ladakh is estimated to about of 500 (Dar 1995). The Bharal or blue sheep (*Pseudovis nahoor*) is a strange looking creature half sheep and half goat. It differs from the goat in the absence of a beard. Likewise it differs from the sheep in the complete absence of glands on the face. The most distinctive external features are the comparatively smooth and olive coloured cylindrical horns, curving upward, down ward, outward, inward and back (Gergan 1962). They have rings denoting the years of growth and from these the age of the animal can be calculated. The weight of a mature ram is about 60 kg. The bharal are found at an altitude of about 6000 meters. In summers they graze in large herds on rich and abundant grasses of meadows.

Ladakh and its neighbourhood is also rich in wild goats. The Asiatic Ibex (*Capra ibex*) is the most beautiful of all. Its sports two long scimitar shaped horns marked with bold transverse knots on the front surface. They may be over a meter long. The mature males have a long beard and weigh over a quintal each. They move in herds of 10-16 and prefer black precipitous rocks and cliffs. In winter they descend to lower attitudes for food and shelter. It is believed that their present number in Ladakh is almost 250.

The Tibetan antelope (Chiru) and gazelle (Goa) are two small but most valuable creatures which inhabit high attitudes in Ladakh. The Chiru (*Pantholops hodgsoni*) is slender built but stout looking due to the thick coat of interwoven pale brown hair. The colouring varies from season to season and from individual to individual. They roam the eastern fringe of Ladakh and inhabit Aksai chin and Tibetan plateau above 5000 meters. Their horns measure upto 70 cm.These animals are slaughtered in large numbers for the sake of underfleece – *Shahtosh* from which the famous *Shahtosh shawls* are made. Hardly a pound of underfleece is obtained from one animal. The Govt. of India has banned sale of Shahtosh shawls. Despite the ban the trade is going on surruptitiously. Unless awareness is created among the buyers there is no possibility of stopping the illegal trade.

The goa or Tibetan gazelle (*Procapra piciticaudatsa*) is more beautiful looking but less common. It is found in Chanthang and Tibet. It lives on the plateaus in the open country and is seldom seen on the mountain sides. It sports horns measuring 36 cm. They roam in herds of 8 to 12.

Among the other herbivores of Ladakh are the mouse hare, Tibetan wooly hare, marmot and many other rodents. Marmots are common rodents found above the tree line. They live in burrows and produce shrill ringing calls. Marmots and the hare are a favorite food item of snow leopard, lynx, fox and large birds of prey. These herbivores hibernate during the winter season.

In Ladakh one comes across the Himalayan bear (*Solenarctos thibatanus*) as well as the brown bear (*Ursus arctos*). The Himalayan bear lives up to 4500 meters while the brown bear lives higher up. The bears are omnivorous animals and feed on anything available plant or animal. According to Dar (1995) there are about 200 each of black and brown bear roaming the heights of Ladakh in summer season. They hibernate during the winter in natural caves or pits dug out by them.

The wolves and the foxes are also found in Ladakh. There are two varieties of wolves. One variety has a light fawn and brown fur while the other is darker. The wolves move in pairs and predate on domestic animals a fact for which they are hated by Ladakhis. However, some of them are trapped and killed for their skins. They have a thick golden fur. They predate on marmots, mouse hare, pheasants and partridges etc.

The most precious predators found in Ladakh are the snow – leopard or Ounce and the Tibetan lynx. The snow leopard (*Uncia uncia*) lives at great heights. It has a soft wooly coat. The upper side has a coat which is grey with cream or buff tinge and the underwool is dark grey or brownish grey. On the body there are irregular large rosettes with dark grey while on the head and neck are solid black spots. The head is clear grey with white around the eyes and upper lip. The backs of the ears are black with a white patch. The fore-paws are spotted while the hind paws a spotless. The snow leopard inhabits rocky hillsides above the tree line. Its natural food is Ibex, shapo, bharal, markhor, marmot and mouse hare. Occasionally it also kills domestic goat, sheep, cattle and ponies of nomads and local Ladakhis. Unless wounded a snow leopard never attacks a man. The skins of ounce fetch fabulous prices a fact which provides enough attention for poachers to kill them. According to Dar (1995) there are around 200 snow leopards still left in Ladakh.

The Tibetan lynx (*Lynx lynx isabellinus*) is an attractive animal. It is found at heights above 4400 meters. It is reported that there are several species of lynx found in J&K state of which the Tibetan lynx inhabits Ladakh. It is a large cat weighing about 25 Kg. The length of the body is 35 to 40 inch and tail is about six inches long. The coat and color vary according to habitat and season. The summer coat is smooth and sleek but the winter coat is long furry with abundance of under wool. It feeds on marmots and mouse hare, snow cock eggs and even young bharal and goat. Sometimes both ounce and lynx steel at night into sheep and goat penns and kill them in dozens by sucking their blood. They are often caught in the process and killed.

The Tibetan wild ass (*Equus kiang*) is a relative of the wild ass of Rann of Kutch. It has wide distribution. Besides Ladakh it is found in Tibet, Turkistan and Mangolia. Kiang is a shy but inquisitive animal. It inhabits plain and rocky grounds and can gallop at a great speed. Solitary animals or groups of 10 to 12 animals can be seen. The colouring of the upper portion of the body of kiang is ruddy chestnut or red and the underparts are white. There is a dark brown dorsal stripe extending from the stiff black mane to the tuft of the tail. The tail has a tuft of black hair.

Avifauna

Despite its inhospitable climate and high and dry terrain Ladakh has a rich and varied avifauna most of which are migrants. There are a few residents too. The most spectacular of them being the snow cocks and partridges. The most well known migrant bird found in Ladakh is the black necked crane (*Grus nigricollis*) called "tung tung" in Ladakhi. These are large graceful birds which inhabit open marsh land, wet plains and prairies. The body is pale grey with black head, neck and feet. The bill is greenish and legs are large. Black necked cranes visit Ladakh during the summer months for nesting. Hussain (1985) has reported sighting of 3 breeding pairs in Ladakh region. The nest is built on a mound in the centre of a marsh from marsh grass and weeds. Two dull white or brown eggs are laid in the nest which is well protected by the parents from predators.

The Himalayan cock (*Tetragallus himalayansis*) inhabits stony slopes between 4000 to 5500 meters. It is a large majestic bird and descends to 2400 meters in extreme cold. These birds are distributed throughout Ladakh and Kashmir but their population has very much declined due to excessive hunting. The Chukor partridge (*Alectoris chukor pallasus*) are colorful birds found widely in Ladakh and Kashmir. They live at a height of above 5000 meters. 5 to 7 eggs are laid in a nest made under the bushes.

Some other notable birds of Ladakh are:

1. Eagle (*Cuncuma leucorypha*)
2. Vulture (*Gyps himalayanus*)
3. Kite (*Milvus migrans lineatus*)
4. Owl (*Tyto alba javanica*)
5. Kestrel (*Falco tinnuculus*)
6. Chugh (*Pyrrhocoax pyrrhocoax*)
7. Mongolian plover (*Charadrius mongolus*)
8. Red Shank (*Triga totanus*)

9. Great Rosefinch (*Carpodacus rubicilla*)
10. Barheaded geese (*Anser indicus*)
11. Black-necked grebe (*Podiceps nigricollis*)
12. Horned lark (*Eremophila alpestris*)
13. Snow pigeon (*Columba leuconota*)
14. Tibetan desert wheateer (*Oenanthe deserti*)
15. Broom headed gull (*Larus brunnicephalus*)
16. Brehimy duck (*Tadorna ferruginea*)

Conclusion

Fauna of Ladakh a trans-Himalayan region of Jammu and Kashmir state is endangered due to human interference and greed. Ever since a network of roads was built in the region and tourist were allowed to come in large numbers the wildlife of Ladakh region has been virtually exposed. Initially armed forces indulged in shikar but are now helping in conservation of wildlife in the region. In the face of the realization that the wildlife in Ladakh region is under threat the Government has come out with measures with the help of both civilians and the army personnel to protect the precious wildlife of Ladakh.

It is relevant to note that because of contiguity of central Asian countries Ladakh, Himachal Pradesh, Uttarakhand, Nepal, Bhutan and Tibet wildlife in these parts is to a great extent homogenous. When a species of wildlife faces threat in one region it flees to the other. That in a way has helped in conservation of these precious species of wild animals in the Himalayas.

References

Dar, Moti, 1995. Wildlife of Ladakh. In: *Ecodegradation of Himalayas*. (Ed.) B.L. Kaul. Vinod Publishers.

Ganhar, J.N., 1979. *The Wildlife of Ladakh*. Hermukh Publication, Srinagar.

Gergan, S.S., 1962. *Big game of Jammu and Kashmir*. A Government of J&K Publication.

Hussain, S.A., 1985. Status of Black necked Crane in Ladakh. *Journal of B.N.H Society*, 82(3).

Kaul, B.L., 1992. Habitat Ecodegradation in Kashmir Himalayas. In: *Conserving Wildlife in India,* (Ed.) S.K. Chadha. Vinod Publishers, Jammu.

Kaul, B.L., 1996. Ecology and status of wildlife in Ladakh. In: *Advances in Fish and Wildlife Ecology and Biology,* Vol. 1, (Ed.) B.L. Kaul, Daya Publishing House, New Delhi.

Kaul, S.C., 1956. *Birds of Kashmir*. Normal Press, Srinagar.

Kendeigh, S.C., 1980. *Ecology with Special Reference to Animals and Man*. Prentice-Hall International.

Chapter 7

Watershed Approach for Integrated National Resources Planning and Management in Himalayas

S.N. Macoo

ABSTRACT

The watershed is a geomorphological unit. It is a drainage area bounded by a divide which contributes run off to a common point or outlet. The area of the watershed is generally determined from the topographic or watershed divide which represents the highest elevation points along the watershed parameter. The present article deals with watersheds of Jammu and Kashmir state which form a part of Indus drainage. These catchments of the state have been delineated and codified. It is now increasingly becoming clear that watershed should be used as a planning unit for rapid, sustainable and integrated regional development.

Introduction

The term 'watershed' refers to the divide which separates one drainage basin from another. The term, is however, generally used to signify a basin or catchment defined as 'land area from which water drains to a given point.' In this context watershed is often synonymous with drainage basin or 'catchment'.

The watershed approach for integrated natural resources planning and management has necessitated a systematic delineation and codification of watersheds at the national as well as state level.

At the national level All India Soil and Land Use Survey Organisation, Government of India, has taken up codification of watersheds at macro-level. Thus the country has been divided into five hierarchical stages as follows:

1. Water Resource Regions
2. Basins
3. Catchments
4. Sub-catchments
5. Watersheds

The entire country has been divided into 6 Water Resource Regions with the following code numbers:

(*a*) Indus drainage = 1
(*b*) Ganges drainage = 2
(*c*) Brahmaputra drainage = 3
(*d*) All drainages flowing into Bay of Bengal except (*b*) and (*c*) = 4
(*e*) All drainages flowing into Arabian sea = 5
(*f*) Ephemeral drainages (namely western desert area) = 6

Indus Drainage

These water resource regions have further been divided into basins, catchments, subcatchments and watersheds. Water resource Region No. 1 (Indus drainage) has accordingly been divided into seven basins, with their code numbers as follows:

Water Resource	Basin	Code No.
1.	Satluj	A
	Beas	B
	Ravi	C
	Chenab	D
	Jhelum	E
	Indus	F
	Ephemeral	G

The rivers Ravi, Chenab, Jhelum and Indus pass through the state of Jammu and Kashmir (Figure 7.1). Codification of watersheds of these basins was taken up by Kango and Qadri way back in 1983 and further carried on by Saran Singh and Gupta (1991-92). The present paper meant to give a summarised account is based on their work spread over several volumes published by the Department of Soil Conservation of Jammu and Kashmir state.

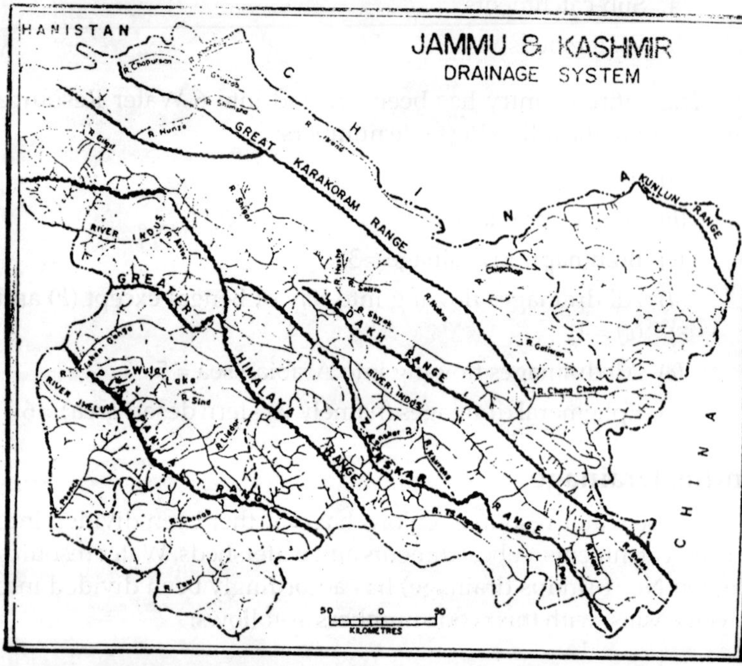

Figure 7.1: Jammu and Kashmir Drainage System

Jhelum Basin (IE)

The basin lies in an oval trough between the Himalayas and Pir Panjal mountain ranges (33°25′ to 34°32′ N and 74°0′ to 75°30′ E) River Jhelum referred to Vitasta in Rigveda is called Veth by Kashmiris. It originates from the spring Verinag and receives a number of tributaries, on its right and left banks during its flow along the middle of Kashmir valley through the alluvium of its own deposition. It falls in the Water Resource Region No. 1 and is further divided into three sub-basins *viz.*, Jhelum (IE), Poonch (IEA) and Kishan Ganga (IEI).

Jhelum Sub-Basin (IE)

Twenty four well defined catchments draining into river Jhelum including terminal catchment covering ephemeral drainage below Baramulla have been identified for the Sub-Basin.There are 60 Sub-catchments, 164 watersheds and 1028 micro-watersheds as shown in the Table 7.1

Table 7.1: Watersheds of Jhelum Sub-basin

Sl.No.	Catchment	Code	Subcatch-ment	Water-sheds	Micro-Water-sheds	Area (in Ha)
1.	Viji-Dhakil	IEU	2	4	22	9,574
2.	Pohm	IEP	8	26	88	1,89,635
3.	Wular I	IEW	2	4	20	14,647
4.	Madhumati	IEM	2	5	43	36,868
5.	Erin	IEE	2	6	33	23,161
6.	Wular II	IEO	2	4	13	8,939
7.	Sindh	IES	8	18	102	1,47,255
8.	Ancher	IEY	2	5	8	8,345
9.	Dal	IEZ	2	4	38	39,500
10.	Arpal	IET	2	5	53	57,100
11.	Liddar	IEL	2	8	98	1,24,300
12.	Kuthar (Arpat)	IEK	2	6	38	36,200
13.	Bringi	IEB	2	7	49	59,200
14.	Sandran	IEC	2	4	26	20,191
15.	Vishav	IEV	2	7	74	98,464
16.	Rambiara	IER	2	6	41	75,100
17.	Romshi	IEX	2	6	23	33,161
18.	Garzan	IEG	2	8	22	12,742
19.	Dudhganga	IED	2	5	49	58,000
20.	Sukhnag	IEQ	2	4	34	41,544
21.	Ferozpur	IEF	2	5	29	35,462
22.	Gunder	IEH	2	4	16	16,195
23.	Nimgal	IEN	2	4	15	14,691
24.	Lower Jhelum	IEJ	2	9	94	1,15,075
	G. Total:		60	164	1028	12,75,696

Poonch Sub Basin

The Poonch river catchment is part of Jhelum basin and lies south of Pir Panchal mountain region, and north of Pakistan. The main Poonch river which originates from Chatta Pani is formed by the subtributaries of Mendhar, Mandi, Suran, Lower Poonch and Hajipeer nallas. These streams have got their sources of glaciers such as sun set peak and other water bodies of Pir Panchal range namely Sannan-Sar, Chinmarg Sar etc. The altitude varies from 2000 M in the North west to 1600 M in the East.

The Poonch river catchment fallings in Jhelum basin (IE) has been divided into five sub-catchments as follows:

1. Mandi river.
2. Suran.
3. Lower Poonch.
4. Mendhar.
5. Hajipeer (not confirmed)

All these rivers join the Poonch river except Mendhar river that joins the stream beyond the actual line of control. These sub-catchments are further divided into watersheds and the code numbers are assigned on the basis of the following principles:

1. The delineation of watershed is done from downstream upwards in a clockwise manner.
2. The codes are allotted adseritum to various systems as they occur along the main drainage line.
3. The systems are delineated on the side of the drainage line.

Each micro watershed is designated by four figure code number (Alal, Ala_2, etc). The first letter designated by a capital English alphabet represents the catchment. The remaining three figures represent the sub-catchments, watersheds and micro-watersheds of catchment respectively. The complete code number for a specific micro-watersheds would thus for example be

$$IEAla_2$$

where,

I: Stands for water Resource Region (Indus)

E: Stands for Jhelum Basin

A: Stands for catchment (Poonch)

J: Stands for sub-catchment (Mandi)

a: Stands for watershed

2: Stands for macro-watershed.

Table 7.2: Watersheds of Poonch Sub-Basin

Sl.No.	*Catchment*	*Code*	*Subcatch-ment*	*Water-sheds*	*Micro-Water-sheds*	*Area (in Ha)*
1.	Poonch	IEA	IEA1 Mandi	4	65	55,670
		IEA_2	Suran River	7	82	59,804
		IEA_3	Lower Poonch	2	18	10,674
		IEA_4	Mendhar	5	55	35,539
		IEA_0	Hajipeer	3	30	31,825
			5	21	250	193512

Krishan Ganga Sub-Basin (IED)

Most of this Sub Basin lies in Pakistan occupied Kashmir and thus could not be codified due to inaccessibility.

Chenab Basin (ID)

River Chenab rises from glaciers on either side of Bara Lacke pass (5724 Mts) in Lahoul in Himachal Pradesh. It flows as two separate streams *viz.* Chandra and Bagha which coverage at Tandi– 30 Km from Thirot in the Pangi valley. The river covers about 130 kms in the valley before entering the hilly terrain of Jammu and Kashmir State. It covers 325 Km of mountainous terrain before descending into plains at Akhnoor. The fall is about 7 mts per Km from Tandi to Kishtwar to Akhnoor below which it becomes navigable. The main tributaries of the river in Himachal Pradesh are Saicuand and Miyar streams. In Jammu and Kashmir the main tributaries are the Bhut, Marusundar, Neeru, Desa, Raggi, Bachelery, Ans and Radh with some ephemeral drainages. The total catchment area of the river is 28,791 sq.Km covering the states of Himachal and Jammu and Kashmir. Codification of Chenab Basin (ID) falling, as it is in Water Resources Region No. 1 has further been divided into various catchments, sub-catchments watershed and micro watersheds as per the Table 7.3.

Table 7.3: Distribution of Watersheds of Chenab Basin

Sl.No.	Catchment	Code	Subcatchment	Watersheds	Micro-Watersheds	Area (in Ha)
1.	Neoshera Tawi	IDO	2	8	120	122124
2.	Munawar Tawi	IDM	3	12	102	163493
3.	Ras	IDR	2	5	46	45079
4.	Ans	IDA	3	11	159	118744
5.	Dhansal	IDI	2	6	30	22544
6.	Mohu-Mangat	IDG	4	11	81	7986
7.	Batote	IDB	2	7	33	33177
8.	Desa	IDD	2	4	31	34113
9.	Kontwara	IDH	3	7	53	47860
10.	Monu-Sudhar	IDW	7	26	353	464919
11.	Kishtwer	IDK	3	13	304	301455
12.	Upper Chenab	IDU	4	H.P. area not attempted	N.A.	781250
13.	Balish	IDS	2	6	190	137709
14.	Niru	IDN	3	10	73	64024
15.	Raggi	IDF	2	5	25	19171
16.	Painthal	IDP	3	7	41	43422
17.	Anji	IDJ	2	5	29	26729
18.	Lower Chenab	IDL	2	8	68	80637
19.	Tawi	IDT	5	16	245	235716
20.	Aik	IDE	2	7	50	57173
				Grand Total		28,79,111

The Codification of Chenab Basin catchments sub-catchments watersheds and micro-watersheds has been done in the same manner as in the case of Jhelum Basin. The complete code number of a specific micro-watershed would thus, for example be

$IDPla_2$:

where,

- I: Stands for water source region (Indus)
- D: Stands for Basin (Chenab)
- P: Stands for catchment (Painthal)
- L: Stands for sub-catchment (Upper Painthal)
- a: Stands for watershed.
- 2: Stands for micro-watershed.

Ravi Basin (IC)

The Ravi known in earlier times as Parushani or Iravati is one of the main tributaries of the Indus Water Resources Region. It rises in the mountain of the Bhangahal basin drainage northern slopes of the Dhauladhar Range and southern slopes of Pir Panjal. It leaves the basin through an inaccessible gorge with perpendicular sides and flows through Chamba in Himachal Pradesh in a north westerly direction parallel to the Dhauladhar Range cutting through that range a few kilometres to the north west of Dalhouse. It leaves the mountain at Basohli after a course of 210 Km during which it drops by 4,570M in altitude (22M per Km). It finally joins the Chinab after running a total course of 720 Km. The Ravi drains three states *viz.* Himachal J&K and Punjab. It forms the eastern and southern boundary of Jammu and Kashmir State with Himachal for about 20 Km and later on with Punjab for about 56 Km.

The main streams contributing the run-off in the Ravi basin in J&K State are Basantar, Tarnah, Ujh, Biani, Chiril and Sewa. The tributaries mostly have stoop slopes and bring flash floods during the rains and remain, day for most part of the year. The total catchment area of the river falling in J&K States is 3,32,395 Ha.

The catchments, sub-catchments watersheds and micro sheds are shown in the Table 7.4.

Indus Basin

The Indus is the westernmost of the Himalayan rivers. It derives its name from the sanskrit Sindhu. The river flows from its origin in Mount Kailash over the Tibetan plateau, through narrow gorges and finally enters the Korakoram range. In Tibet the river is called Singhchu (Singh-Lion; Chu- Mouth) as it is believed to originate from the mouth of a lion. It is one of the mightiest rivers of the world, draining the glaciers and mountain slopes of many famous peeks like Aling Kangri (7,315m), Tirich Mir (7,690 m) Gasherbrum (8008 m), Masherbrum (7,821 m), K2 (8,610 m) Rakaposhi (7,788 m) Nanga Parbat (8,114 m) etc., and receiving a galaxy of great rivers as tributaries. The total length of the river is 2,900 km and the drainage basin is estimated to have an area of 96,600 sq.km. It flows for about 290 km over a flat country along the northern (inner) flank of the Ladakh Range. Then it cuts across that range at Thangra and flows along the outer flank for another 480 km, near Skerdu (altitude 2,610 m). It cuts the Ladakh range again, resuming the general trend of its course on the other side. After circling round the Nanga Parbat (8,114 m) it flows south-west through Hazara towards the plains of the western Punjab.

Table 7.4: Distribution of Ravi-Basin Watersheds

Sl.No.	*Catchment*	*Code*	*Subcatchment*	*Watersheds*	*Micro-Watersheds*	*Area (in Ha)*
1.	Basanter	ICB	4	10	59	50,800
2.	Ujh	ICU	5	18	162	1,34,120
3.	Sewa	ICS	2	7	47	43,200
4.	Chiril	ICC	3	6	24	19.450
5.	Biani	ICD	2	6	46	36,875
6.	Tarnah	ICT	2	7	49	47,950
				Total		3,32,395

The chief tributaries of the Indus are the Zanaskar which rises on the Zanaskar range, cuts through it and joins below Leh; the Drass river which comes from the northern side of Zojila pass and the plains of Deosai; the shyok which rises on the northern side of the Karakoram Range and cuts across the south and joins the Indus near Kiris; the Shigar which drains the southern slopes of K2 and Biafo and Baltoro glaciers; the Gilgit (with its Hunza tributary) which rises behind then cuts through the Korakoram Range.

The Indus is a large river over 150 m wide and 3 m deep at Skardu, even in winter. It passes through gut in chitral and then traverses 200 km of the wild territory of Kohistan entering the N.W. frontier province near Darband. At Attook 11,385 km from the source it receives the Kabul tributary flowing in from Afghanistan and then flows due South. Here the bed of the Indus is 610 m above the sea level and its width varies from 90 to 230 m according to the season. Below Attock the Harok and Sohan (Soan) rivers join the Indus. At the confluence of the Punjab rivers near Mothankot the river is 350 m wide and its normal discharge is 25,000 cusecs which increases in flood time to about million cusecs. This confluence is 85 m above sea level and 790 km from the sea. The river sometimes rises as much as to 6 to 7 m in flood. The delta, which is 7,800 sq. km in extent, contains several old channels and is entirely flat with practically no large trees.

26 catchments have been identified and delineated as per the methodology adopted for codification of Indus Basin falling in Ladakh region of J&K State (Table 7.5).

Table 7.5: Distribution of Catchments of Indus Basin (IF)

Sl.No.	*Name*	*Code*
1.	Damker	IFY
2.	Limlahfu	IFU
3.	Leh	IFC
4.	Lalungo	IFL
5.	Tangtse	IFA
6.	Shyok	IFB
7.	Diskit	IFC
8.	Turtolangpo	IFT
9.	Fastaulangpo	IFF
10.	Nobra	IFN
11.	Rangelologpo	IFR
12.	Shukpokunchung	IFS
13.	Ororotse	IFO
14.	Dangoyalko	IFG
15.		IFI
16.		IFJ
17.	Damchok	IFU
18.	Henle	IFH

Contd...

Table 7.5–Contd...

Sl.No.	Name	Code
19.	Tsomorari	IFP
20.	Niritospochu	IFQ
21.	Upshi	IFV
22.	Zanskar	IFZ
23.	Sura	IFW
24.	Mirdha	IFM
25.	Sankpo	IFX
26.	Drass	IFD

Drass-Sankpo River System

Out of 26 catchments only two well defined catchments of Drass and Sankpo river are described here.

The Drass river is formed of the united streams of the Wakha, the Suri, the Drass and the Kaksar rivers. It originates from Zojila pass and forms the catchment of Drass river. The other river system of Sankho, drainage joins the Drass river at Bethong which subsequently joins "Sengge chhu "at Olking Chung approximately 20 Km away from the actual line of control.

The Drass river system with river order of Sanphoo lies in a oval trough between Himalaya and Korakaram mountain ranges (33°–47′ to 34°–46′ North and 75°–16′ to 76°32′–15′E). The system passing largely along the middle of the Drass Valley through recent allurum of its own deposition, is the recipient of the drainage of Drass Valley. The Drass river and Sankpo river are one of the catchments of 'Senggechhu". The river Sindh among Drass and Sankpo rivers, receives a number of tributaries on its left and right banks. These include fairly developed streams as well as tiny rivulets with quite remarkable variations in their drainage and catchment characteristics.

The methodology used in codification of the Indus is based on the following principles:

1. The delineation of water shed start from down stream upwards in clockwise manner.
2. The codes are allotted ad-seritum to various systems as they occur along the main drainage line.

3. The system are delineated on one side of the drainage line. Transverse or cross cutting on drainage line is resorted to in the case of elongated drainages and only in exceptional cases.

Each micro-watershed is designated by four figure code number (Dlcl, D_2cl and D_2 b_4 etc.), the first letter designated by a capital English alphabet represents the catchment and corresponds invariably with the first letter of the name of the river covering the catchments.

The remaining three figures represent the sub-catchments, watersheds and micro-watersheds of the catchment respectively. The complete code number of a specific micro-watershed would thus for example be

IFDla_2.

where,

I: Stands for water resource region (Indus)

F: Stands for Basin (Indus)

I: Stands for catchment (Drass)

a: Stands for watershed

2: Stands for micro-watershed.

Drass-Sankho catchments have 10 sub-catchments, 35 watersheds and 497 micro-watersheds covering an area of 6,51,038 Ha (Table 7.6)

Conclusion

Watershed management has been defined as "The analysis, protection, repair and utilization of drainage basins for optimum control and conservation of water with due regard to other resources." The definition summarises the tasks of watershed manager and the objectives of management. The watershed management is a conscious effort of selecting and developing the best way of managing the renewable natural resources. According to the procedures laid down by Govt. of India. "Watershed is an integration of technologies within the material boundaries of drainage area for optimum development of land, water and plant resources to meet the basic minimum needs of the people in a sustained manner. Projectisation of the scattered programmes of soil conservation, afforestation, minor irrigation, animal husbandry and others offer hope for bringing about sustainable natural resources development based on principles of ecology, economics, employment generation and energy conservation." Watershed management is thus the development and management of watershed resources to achieve optimum production which could be sustained without deterioration of the resource base or disturbing the balance of regional ecological environment. Watershed may be divided into macro, micro and mini watershed for management purposes. But the highest priority in watershed management should be given to initiation of progammes for restoration of optimum hydrological functioning of the watershed.

Table 7.6: Distribution of Indus Basin (Drass Sankho Catchments)

Sl.No.	Catchment	Subcatchment	Code	Watersheds	Micro-Watersheds	Area (in Ha)
1.	Drass	Drass	IFDI	3	28	36,747
		Gumri	IFD2	4	37	57,913
		Mashko	IFD3	4	41	43,750
		Shingo	IFD4	3	44	52,319
2.	Sanpho	Workharang	IFX1	3	100	1,07,371
		Sanpho	IFX2	4	66	1,03,655
		Karchoo	IFX3	3	32	46,874
		Barso	IFX4	4	50	62,050
		Shangra	IFX5	3	49	70,979
		Kargil	IFX6	4	50	69,380
			G. Total	35	497	6,51,038

In terms of resources development, it starts from the most important one, that is water and land (land use and cropping pattern) and their characteristics and extends to the resources of fuel, fodder, fiber, livestock and all associated components.

For better management of watersheds and an integrated natural resources planning delineation and codification of watersheds at the National as well as State level has become necessary. At the National level All India Soil and Landuse Survey Organization Govt.

of India has taken up codification of watersheds at macro-level. The present article deals with codification of watersheds of Jammu and Kashmir and will help watershed managers in utilization of water resources of the State, in a more useful manner.

It is important to note that a watershed as a basin of a river has a physico-eco-cultural homogeneity. Any planning in the hills is only successful when it is planned at a watershed level, because it constitutes a technical and ecological unit which is self-contained physically, composite functionally and coherent for land and water resource use planning and management. It is a naturally drawn physico-geographical entity in which an integrated strategy is planned for land utilization, irrigation, flood control, power generation, industrial and domestic water supply, soil and water resources conservation and effective implementation. The approach provides the best way to manage the present and future needs of the community without jeopardizing or destroying the ecological balance. Thus production becomes resource centred and environment based and helps in promoting a sustained process of rapid, balanced and integrated regional development.

Chapter 8

Dal Lake: A Shrinking Himalayan Jewel

B.L. Kaul

ABSTRACT

Dal lake in Kashmir Himalayas as per historical records was about 76 sq. Km in 15th century but was reduced to 25 sq. km in area a hundred years ago. Today it has been reduced to less than 10 sq. km in area. The chief causes of shrinkage of the lake have been inflow and accumulation of inorganic sediments, plant nutrients and human activity in and around the lake. The lake will disappear in another 50 years if remedial measures on a war footing are not taken.

Introduction

The earliest reference to Dal lake is available in Nilmata Purana (6th Century A. D.) wherein it is referred to as "Sureshwari" whose water had healing powers. Kalhana (12th Century A. D.) in "Rajatarangini" calls it "Jewel of a lake". He writes that Dal lake extended from Shankarkcharya hillock in the south to Mahadev mountain in the north and from the present day Sodarbal in the east to Telbal in the west. Srivara (15th Century A. D.) writes in his teatise "Rajvalipataka" that the earlier name of Dal lake was "Sureshmari" which was later on changed to "Sureshwari". The term Dal seems to have come into use sometime later. Srivara further writes that Sultan Zain-ul-Abidin (1420-70 A. D.) often used to visit Dal lake in a boat.

According to this chronicler the lake extended upto a length of 12 miles (19 km) from South to North. It is further recorded that the lake extended upto Sidhar Puri (Soura) Zainanagri (Nowshehra) and Zushkpur (Zokur) on the western side and in the north it extended upto Burja Dhama (Burzhom – the famed site of prehistoric pitdwellers). There is mention of floating gardens (called Radh in Kashmiri) neither in "Rajatarangini" nor in "Rajvalipataka". It is believed that floating gardens – a very important source of vegetables to Srinagar city and other towns – were introduced in Dal lake during the Mughal rule.

In comparatively recent times Fredrick Drew (1875), Lawrence (1895) and Stein (1900) also made observations about the Dal lake during their stay in the valley of Kashmir. There are minor variations in their observations regarding the size of the lake. While Drew has stated that it was 5 miles (8 km) in length from North to South and 2 miles (3.6 km) in width from east to west, Lawrence and Stein have stated that it was about 4 miles (6.4 km) in length and 2.5 miles (4 km) in width. Despite these variations it can be safely presumed that by the turn of this century Dal lake had an area of about 25 sq. km. The maximum depth of the lake observed around this time was 30 ft. Mukherjee in 1930 recorded that Dal lake had a maximum depth of 20 ft. In 1964 the open water area of Dal lake had shrunk to 13.25 sq. kms and in 1980 it was a mere 11.7 sq. km. The average depth observed around this time was 2 meters.

While increase in agricultural activity in and around Dal lake has directly and indirectly contributed to its shrinkage and present state of eutrophication, there are other and more important causes. The main cause of the shrinkage of Dal lake is the continuous inflow of inorganic sediments plant foods and nutrients from the surrounding catchment areas and accumulation of insoluble and soluble materials. The often unauthorized and disorderly growth of human settlements in and around the lake and flow of effluents and wastes from house boats, hotels and mushrooming guest houses have all added to the pollution of the lake resulting ultimately in its shrinkage.

The situation today is alarming despite a "Dal Development Authority" in place. It calls for immediate adoption of remedial measures on a war footing to save this beautiful lake from certain death.

Table 8.1: Area of Dal Lake in sq. km in Recorded History

Sl.No.	*Period*	*Reference*	*Name*	*Size*	*Open water Area*
1.	6th Century A.D.	Nilmata Purana	Sureshwari	Not Mentioned	Not Mentioned
2.	12th Century A.D.	Kalhana's Rajataringini	–Do–	–Do–	–Do–
3.	15th Century A.D.	Srivara's Rajavalipataka	Sureshwari	19 x 4 km	76 sq. km
4.	19th Century A.D.	Lawrence (1895)	Dal Lake	6.4 x 4 km	25.6 sq. km
5.	–Do–	Stein (1900)	–Do–	–Do–	–Do–
6.	20th Century A.D.	Mukerji (1930)	–Do–	5 x 4 km	20 sq. km
7.	–Do–	Survey (1964)	–Do–	–	13.25 sq. km
8.	–Do–	Survey (1980)	–Do–	–	11.7 sq. km
9.	21st Century A.D.	Survey (2001)	–Do–		10 sq. km

Physical Features

Dal lake comprises of five sub-lakes or basins and many inter-connecting channels. This five basins are :-

(*i*) The Hazratbal

It constitutes the major part of the lake being 8.8 sq. km in area (5.6 sq. km of open water and 3.2 sq. km of marsh).

(*ii*) Bod Dal

It consists of 4.2 sq. km of completely open water.

(*iii*) Gagribal

It comprises of 1.3 sq. km of open water.

(*iv*) Floating Gardens

These comprise of 4.8 sq. km area with 4.5 sq. km of marsh and only 0.3 km of clear water.

(*v*) Boulevard Area

It consists of 0.5 sq. km of which 0.3 sq. km is open water and 0.2 sq. km of marsh.

Thus Dal lake has as per a survey conducted in 1980 a total area of 19.6 sq. km of which only 11.7 sq. km is open water area and the rest is marsh.

The most important stream draining into Dal lake in the north is the Telbal Nullah. It brings water from the high altitude lake Marsar and flows through the large Dachigam valley before it drains into the lake. Several small streams around the shores of Dal lake and springs under the lake bed also feed it.

On the south west corner of the lake lies the Dal gate which has a weir and lock and drains into Chinar Bagh – a loop channel of river Jhelum. A part of the lake water drains via Nagin Lake and Nullah Amirkhan to Anchar Lake. Dal lake was also previously drained partly by Nullah Mar via Brari – numbal a shallow basin into which effluents and sewage from parts of the city flow. Some years back the Nullah Mar which drained Brari – numbal into river Jhelum was filled and converted into a road. This unmindful step has added to the pollution of Dal lake because the sewage and effluents that collect in Brari – numbal basin are finding their way into the lake now.

The Boulevard road borders the eastern and southern sides of the lake. It starts from Dal gate and loops around Gagribal and Bod Dal.

Catchment Areas

The catchments of Dal lake are Dachigam, Telbal, lake hillside and Srinagar city. The total catchment area contributing to lake inflows is about 316 sq. km. Dachigam and Telbal catchments consist of mountain ranges and cultivable land. The lake hillside has very little vegetation. The famous Mughal gardens overlooking the lake are situated on the hillside. The gently sloping terraces around the lake are extensively used for paddy cultivation, orchards (of apple, pear, cherry, apricot, peach, plum, almonds and walnuts) and vegetable gardens.

Table 8.2: Area of Dal lake Catchment in 1989

Sl.No.	*Name of Catchment*	*Area in sq. km*
1.	Dachhigam	148
2.	Telbal	80
3.	Lake hillside	47
4.	Srinagar Dal catchment	21
5.	Dal lake	20 including marsh

Table 8.3: Inflow of Plant Nutrients into Dal lake

Sl.No.	*Nutrient*	*Inflow per Year*	*Estimated Deposits in Upper 20 cm*
1.	Nitrogen	3221.1 tonnes	7,722 tonnes
2.	Phosphorus	14.8 tonnes	3,744 tonnes

Geological Features

The chief geological formations of the catchment area of Dal lake are panjal trap, lime bearing rock (Marl), agglomerate slate and clay. Freshly laid alluvium is also present in parts of the lake. Leeching out of calcium from the catchment makes the lake water hard. The bottom is thickly covered by alluvium.

It was till recently held that the Dal lake is a remnant of a glacial lake which existed during Pleistocene period. However, remote sensing studies have shown that the lake has been derived from enlarged oxbows and abandoned flowed channels of river Jhelum.

Thermal Stratification

Dal lake being shallow with temperature ranging between 5°C in winter to 30°C in summer, it does not undergo thermal stratification. The solubility of oxygen in water decreases at higher temperature. Even though water temperature in summer rises to 30°C, high level of photosynthetic activity within the lake results in rise of D. O. level of the surface water to super saturation values upto 19 mg/l the lower limit being 7 mg/l. The bottom water in Hazratbal and Gagribal basins are also well oxygenated (min. D. O. 5.2 mg/ l) throughout the year while in the nearby Nagin lake, deoxygenation of the hypolimnion (lower layer) occurs in summer (min. D. O. 0.2 mg/l). This level is not conducive for aquatic animal life.

Biological Characteristics

The biological state of the Dal lake is such that it has mixed flora dominated by diatoms and followed by blue green algae. The population density to phytoplankton is low. Copepods, rotifers and protozoans are the principal components of the zooplankton which are low in number.

There is extensive growth of macrophytes and water fern covering most of the lake floor thereby reducing the aesthetic quality of water. The common fish are the exotic mirror carp and scale carp which are able to live in waters containing very little oxygen. Being mainly herbivorous these fish are able to get abundant food from the lake vegetation. Prior to the introduction of these exotic fish in Dal lake the endemic fauna consisting mainly of various species of *Schizothorax* flourished. Now their populations are dwindling due to pollution of the lake and also the competition with the more hardy mirror carp and scale carp.

Why Dal lake is Shrinking?

There is no doubt about the fact that Dal lake is shrinking and that it has become increasingly marshy. From having an open water area of about 25 sq. km in 1900 it had shrunk to an open water area of 11.7 sq. km only in 1980. The main causes of its shrinkage are discussed here.

Table 8.4: Biological Status of Dal lake

Sl.No.	Major Life Forms	Types
1.	Macrophytes	*Typha, Hydrilla, Nelumbium, Nymphaea, Vallisnaria, Potamogeton, Salvinia, Phragmites, Myriophyllum, Limnamthemum, Utricularia*
2.	Phytoplankton	Diatoms, Blue-green algae
3.	Zooplankton	Protozoans, Rotifers, Copepods, Insect Larvae
4.	Fish	*Cyprinus, Schizothorax, Botia, Gambusia, Crossocheilus, Barbus, Labeo, Nemacheilus*
5.	Bacteria	Coliform Bacteria, Streptococci

Inflow and Accumulation of Inorganic Sediments and Plant Nutrients

Studies have shown the presence of soluble Ca, Mg, K, P and N in high concentration in the lake water as we pass from the mouth of the Telbal Nullah towards the center. The average annual nutrient accumulation has been calculated to be 14.8 tonnes of phosphorus and 322.1 tonnes of nitrogen. It is believed that 2.05 tonnes of phosphorus and 143.5 tonnes of nitrogen are incorporated in the weed annually. Some of the weed is extricated by the owners of floating gardens and added as a manure to the vegetables grown by them. It is estimated that 3.744 tonnes of phosphorous and 7.722 tonnes of nitrogen has accumulated in the upper 20 cm of sediment, over the years, which is available to the plants growing in the lake. The presence of excessive plant nutrients has promoted rooted weed growth which now constitutes an important factor for the present and future changes of the lake ecosystem.

The alkaline character of the lake water (pH ranging from 7.4 to 9.5) rich in calcium is responsible for the precipitation of phosphorus and deposition in the sediments. This results in relatively low phosphorus content in lake water and comparatively rich content in the sediments. High pH is harmful for most aquatic animals and results in formation of restricted habitat.

Soluble nutrients also cause phytoplankton and floating weeds *e.g. Salvinia* to grow, while insoluble nutrients bound in sediments of the lake bed provide food for the rooted weeds (macrophytes). Considerable quantities of floating weeds in the back waters of the lake and profuse growth of rooted weed all over the lake area are quite evident.

The insoluble sediments in the form of silt are brought in by the inflow channel. In recent years the clarity of Dal lake waters has lessened due to the presence of large quantities of insoluble particles and deposition of sediment in the lake bottom has made it shallower.

Agricultural Activity

Land use in the catchment areas has played an important role in the deterioration and degradation of the catchments resulting in accelerated rate of soil erosion and consequent pollution of Dal lake. The agricultural activity on reclaimed land and on the floating gardens have also contributed to the pollution of the lake. The floating gardens are the chief source of vegetables for the city of Srinagar. Boat loads of vegetables are ferried down Dal gate to different ' ghats ' on the river Jhelum every day in summer. Succulent lotus stems grown in Dal called " Nadroo " are considered a delicacy by the local populace. The quality of " Nadroo " from Dal is certainly better than those grown in Anchar and Wular Lakes. The achenial fruits of the lotus are relished when fresh. These are also used in a dried state for performing Yagneas by Hindus. Large lotus leaves are occasionally used for serving food in the same manner as banana leaves are used in southern parts of the country. Other aquatic plants cultivated in back waters of Dal are " Sanghara " (*Trapa bispinosa*) and water lily. While sanghara seeds are eaten in a number of ways, flour obtained from dry seeds is used as an alternate food for rice by diabetics.

The reclaimed land is also being used for growing vegetables which are supplied to the city population besides the populations of important towns like Baramulla, Sopore and Anantnag.

The terraced hill side of the lake is extensively used for paddy cultivation. Maize and wheat are also grown on a smaller scale. These gently sloping terraces are ideal for horticulture. Thus there can be seen large areas which have been converted into orchards. Even the Mughals were attracted by the beauty of the lake hillside to construct the famous "Mughal gardens" and "Pari Mahal".

The main catchment areas of the Dal are the Dachigam and Telbel. These consist of mountain ranges and cultivable land. There has been considerable encroachment of forest land for cultivation which has further added to soil erosion.

Due to modernizing in agricultural operations and indiscriminate use of insecticides, fungicides, fertilizers, growth regulators etc. there has been an overall deterioration of the lake ecosystem. These substances invariably find their way during rains into Dal lake and affect biological activity. Insecticides like D. D. T. enter food chains and reach humans and domestic animals. The fish of economic value which are found in the Dal lake include species of the endemic *Schizothorax* and the exotic mirror carp (*Cyprinus carpio*). Prior to introduction of the mirror carp in Kashmir waters in 1955 the Dal lake used to have large populations of *Schizothorax*. The size attained by these fish used to be large. The situation today is totally different. In every catch the fishermen hardly get 10 per cent of small sized *Schizothorax* while the rest are mirror carp. The reasons for this situation are not difficult to find. Continuous degradation of the quality of water has made the Dal lake unfit for *Schizothorax* to live in. On the contrary the mirror carp is able to live in waters containing low oxygen and being mainly herbivorous it finds enough food from the lake vegetation.

Land Reclamation and Human Settlements

In recent past a lot of human activity such as land reclamation, encroachments and growth of unauthorized and disorderly human habitations and construction of hotels and guest houses for tourists around the lake have contributed to its pollution and resulted in further deterioration of the water quality. Consequently the change in the ecosystem in such surroundings has defeated the very purpose of the beneficial use of the water.

Overgrazing of the Catchment

Uncontrolled grazing in the catchment is causing erosion of top soil. Deforestation and sliding of steep hillside slopes during torrents further add to the problem and result in sediment inflows in the lake. It is estimated that sediment inflow due to overgrazing and soil erosion is about 46 acre feet per annum during normal years and 64 acre feet during the flood years.

The denudation of hill slopes, hardening of the soil by overgrazing and destruction of the vegetal cover in the catchment areas are responsible for excessive run off and bringing down of the soil and detritus which is responsible to a great extent for silting up of Dal lake.

Flow of Effluents and Wastes into Lake Waters

As mentioned earlier some of the domestic effluents and wastes are drained into batckwaters and channels from parts of Srinagar catchment area of the Dal lake. Mushroom growth of hotels and guest houses have further added to this load. The house boats for which Dal lake is famous are fitted with sanitary fittings and flush latrines. In the absence of a sewerage system for the house boats effluents and sewage from them are being discharged into the lake waters. This has drastically changed character of the lake water as is indicated by the presence of *coliform* group of bacteria and faecal *streptococci* especially in Hazaratbal and Boulevard basins. The coliform bacterial count in Hazaratbal and Boulevard basins is 150 MPN/100 ml and 2400 MPN/100 ml respectively. In Bod Dal and Gagribal basins the count is comparatively low being 640 MPN/ 100 ml and 750 MPN/100 ml respectively. The high count in respect of Boulevard basin is due to flushing of house boat latrines into the lake waters.

What is the Remedy?

Fortunately the state of degradation of the Dal lake has not yet reached the point of no return. This fact has gone home and the government of Jammu and Kashmir have initiated steps to save this beautiful lake from further deterioration. A "Dal Development Authority" has been formed to reverse the trend. It envisages the following steps:

Settling Basin

In order to check the inputs of sediments and nutrients from the catchment area a settling basin is being constructed at the mouth of the Telbal nullah.

Foreshore Road

A foreshore road is being constructed around the lake to stop encroachments and unauthorized construction of hotel buildings and guest houses. Parts of this road have yet to be constructed.

Dredging and Deweeding

Dredging of the lake in order to remove the sediment and increase the depth is in progress. It will require enormous effort and time to dredge the lake bottom. Deweeding, by machines, however, has been undertaken in order to get rid of the submerged and floating vegetation which has been affecting the quality of the water.

Sewerage of House Boats

An extensive system of sewerage of the house boats has got to be made in order to prevent pollution of the lake water from the faecal matter and other effluents coming out of them.

Sewerage System for Srinagar City

In order to improve the water circulation of the lake and prevent its pollution by the sewerage from the Srinagar catchment area, construction of a sewerage plant near Brarinumbal has been conceived. UNESCO has agreed to provide funds for the construction of a sewerage system for the Srinagar city. If and when constructed there will be a considerable improvement in the quality of Dal waters by diversion of the movement of the city effluents and increased water flow of the lake.

Restricting Agricultural Activity

The agricultural activities have got to be restricted to permissible grades of slopes in the catchment areas. Encroachments of steep forest lands for agricultural activity will have to be stopped. Use of chemical fertilizers, insecticides, fungicides etc will also have to be restricted.

Afforestation

There is reason to believe that in not so distant past hillside of the lake was full of vegetal cover. It is completely denuded now. An extensive programme of afforestation of this hill range will have to be undertaken. Besides this Dachhigam and Telbel catchment areas too will have to be covered under a programme of afforestation on a war footing. This single step will prevent soil erosion and considerably check flow of sediment and nutrients into the lake and there will be a visible restoration of the lake ecosystem.

Regulation of Grazing

Improvement of pasture land and regulation of grazing in the catchment area are other important steps which can help in calling a halt to the process of shrinkage of the lake. Pasture lands can be made protective by reseeding with economic grasses and legumes combined with controlled grazing. An appropriate grazing policy is required to be laid down for the catchment areas of Dal Lake to stop its further degradation.

References

Dhar, K.N., 1983. Dal Lake through the ages. In: *Abstracts, National Workshop on Conservation of Dal Lake.* University of Kashmir.

Drew, F., 1875. *The Jammu and Kashmir Territories.*

Kalhana (12th Century A. D.). *Rajataringini.*

Kaul, B.L., 1992. Eco-degradation of Dal Lake. In: *Himalayan Environment Man and the Economic Activities,* (Ed.) J.L. Raina. Pointer Publishers, Jaipur.

Lawrence, W., 1895. *The valley of Kashmir.*

Srivara (15th Century A. D.). *Rajvalipataka.*

Stein, A., 1900. Chronicles of the Kings of Kashmir. English Translation of Rajatarangini, Vols. I and II.

Verma, N.K., 1983. A review paper on effect of pollution on Dal Lake. (Unpublished)

Zutshi, D.P. and Mufti, N.A., 1983. Evaluation of Trophic status of two interconnected basins of Dal Lake. In: *Abstracts, National Workshop on Conservation of Dal Lake.* University of Kashmir.

Zutshi, D.P. *et al.*, 1983. Modification in physical and chemical environment of Dal Lake. In: *Abstracts, National Workshop on Conservation of Dal Lake.* University of Kashmir.

Chapter 9

Conservation of Medicinal Plant Resources in Western Himalaya

O.P. Sharma

ABSTRACT

Western Himalayas is a home for rich medicinal flora. Out of 6000 diverse kinds of species, 1700 plants are used as source of drugs or folk medicine. As concern for environmental degradation has become a global concern, the need for conservation action has become more important. What species, how and where need to be conserved, need some kind of threat assessment. Based on IUCN Red list criteria's species have been categorized as Extinct, Extinct in the wild, Critically Endangered, Vulnerable, Near Threatened, Least concerned and Data Deficient. Prioritization of plant species is going to help in effecting *in situ* and *ex situ* conservation programmes for conserving rich floral wealth for our future use.

Keywords: *Red data lists, IUCN, FRLHT, Critically endangered, Endangered, Vulnerable.*

India is home to well known traditional medicine systems that support many of the healthcare needs of the masses. Healing properties of medicinal plants stands discussed in ancient Ayurveda texts. The so called Indigenous Systems of medicine comprising of Ayurveda, Unani, Siddha, Homeopathy and Tibetan are almost entirely based on the utilization of medicinal plants and their derivatives in raw or processed form for prevention and treatment of various ailments and disorders.

Western Himalayas is quite rich in medicinal plant resources. Out of 17500 known flowering plants of India, more than 6000 plants are found in this region. Out of 6000 west Himalayan plant species, 1700 are known medicinal plants which find refuge in this area. Habit-wise breakup of the medicinal plants indicates that the highest proportion is of herbs (about 42 per cent), followed by trees (about 25 per cent), shrubs (about 18 per cent) and climbers (15 per cent). Variety of botanical drugs obtained are mostly from roots followed by whole plant, fruit and seed, leaves, bark, flower, stem, gum etc.

A report on demand study for selected medicinal plants by Centre for Research and Action (Anon, 2000) has reported that about 150 plants account for major share (about 70 per cent) in the domestic trade. The highest recorded demand is for *Emblica officinalis* (Amla, more than 15000 tonnes) followed *Asparagus racemosus* (Shataver, more than 8000 tonnes), *Withania somnifera* (Asgandh, more than 6000 tonnes), *Terminalia chebula* (Harad, more than 5500 tonnes), *Aegle marmelos* (Bill, more than 4400 tonnes), *Adhatoda vasica* (Adusa, more than 4200 tonnes). Other plants which are in the trade include *Sida cordifolia* (Bala), *Woodfordia fruticosa* (Dhataki), *Boerhavia diffusa* (Punamava), *Picrorhiza hurroa* (Katuki), *Swertia chirayita* (Charaiyita) *Valeriana jatamansi* (Mushakbala), *Pistacia integerrina* (Kakar Singhi), *Jurinea macrocephala* (Dhoop), *Angelica glauca* (Choru) etc. *Saussurea costus* locally called Kuth (Minaal in Shina) is confined to the Kashmir Himalayas. Kuth Act was promulgated in J&K during Maharaja's time. Under Kuth Act not only Kuth (*Saussurea costus*) but other plants like *Artemisia brevifolia* syn. *A. maritima* (Vern. Morri, Seski), *Podophyllum hexandrum* syn. *P. emodi* (Vern. Bankakkri, Denmokushu), *Atropa acuminata* (Vern. Tookansool, Jalakaffal) also got included. Upto 1947, Kuth cultivation was conspicuous in J&K and it had been going on a large scale especially in Gurez and Astore area (now under Pak occupied Kashmir). Kuth cultivation was also undertaken in and around Simthan pass in the past. From times immemorial, China has been our biggest importer of Kuth and Kuth was the major revenue earner for the state of Jammu and Kashmir. Although Kuth cultivation was abandoned in J&K, thanks to Lahul-Spiti farmers who are still growing and exporting Kuth to China, Japan, Korea Singapore and Thailand. Kuth Act of J&K stands repealed now but many of these medicinal plants find mention in the Schedule VI of J&K Wildlife Protection Act, 2004.

Medicinal Trees

Many trees of Sub-tropical dry deciduous and moist forests are known for their therapeutic value. Three myrobalans, embelic, chebulic and belleric myrobalans are abundant in the Kandi belt of J&K and also seen in H.P. *Terminalia arjuna* (Arjun), *Moringa oleifera* (Sohanjana), *Dillenia indica* (Chalta), *Aegle marmelos* (Bill, Vilva), *Butea monosperma* (Palash), *Erythrina indica* (Thuba), *Holarhaena pubescens* (Kogar), are all medicinal trees of Shiwaliks. Our hills and mountains support the growth of many medicinal trees like *Aesculus indica, Bauhinia variegata, Prunus cerasoides* etc.

Medicinal Shrubs

Sub-tropical area is rich in many medicinal shrubs like *Justicia adhatoda, Barleria prionitis, Barleria dichotoma, Vitex negundo, Jatropha curcas.* Mountain medicinal shrubs seen in western Himalayas are *Osyris orborea, Piptanthus nepalensis, Desmodium elegans, Reinwardtia indica, Prinsepia utilis, Artemisia brevifolia.* Cold desert area of Ladakh and Lahul Spiti is rich in *Ephedra gerardiana* (Chhepath) and Hippophae rhamnoides (Sarbang).

Medicinal Herbs

Many medicinal herbs are exploited for their therapeutic and healing properties. Sub-tropical herbs include *Ocimum americanum, Hyptis suaveolens, Psoralea corylifolia, Salvia plebia, Micromeria biflora, Acorus calamus, Eremostachys superba.* Our temperate forests and sub-alpine meadows are rich in medicinal herbs like *Jurinea macrocephala* (Dhoop), *Armebia benthami* (Kah-Ajwain), *Swertia petiolata* (Chirayta), *Picrorhiza kurroa* (Katuki), *Tussilago farfara* (Chichiloti), *Geranium wallichianum* (Rooli). Cold desert of Ladakh is quite rich in medicinal plants like *Aconitum heterophyllum* (Buma), *Allium carolinianum, Nepeta floccosa, Codonopsis clematidea* (Bruktung), *Tanacetum tibeticum, Physochalaina praelta* (Langtang), *Artemisia stracheyii, Waldheinia tomentosa* (Nakpo), *Corydalis flabellata* (Nyatpo). *Dactylorhiza hatagirea* (Salampanja).

Medicinal Climbers

Sub-tropical belt in J&K and H.P. supports the growth of many medicinal climbers. *Tinospora cordifolia* (Giloe), *Pueraria tuberosa*

(Badaar), *Cissampelos pariera* (Patha), *Diplocyclos palmatus* (Shivlingi), *Celastrus paniculatus* (Malkangni), *Gloriosa superba* (Langli). Our mountain forests have some medicinal climbers like *Marsdenia roylei, Clematis tibetana, Cuscuta europea, Ipomoea muricata* which are used by folk healers as native medicine.

Medicinal Orchids

The Aesthetic value of orchids is world famous. The most revered Shri Mata Vaishno Devi Shrine is decorated with exotic orchids during Navaratras. Orchids have medicinal value in addition to decoration. Saalam panja of our temperate forests is an orchid, *Dactylorhiza hatagirea* (Marsh orchid, Palm orchid), whose roots are medicinal and used as aphrodisiac. *Habenaria plantagenia* found in Dalhousie and Chamba forests is also medicinal and Koya tribes employ this species in their folk treatments. *Nervilia plicata* grows in Himalayas including the Trikuta hills and has medicinal value. Root preparation of *Dendrobium nobile* (Vern. Jeevaki) is used for the treatment of general debility. It is also used in the Chyawanprash. *Eulophia campestris, Satyrium nepalense, Zeuxine stratematica, Habenaria commelinifolia, Habenaria intermedia, Habenaria grandiflora, Cypripedium cordigerum, Malaxis muscifera* etc. are other medicinal orchids found in Western Himalayas. They possess antiseptic and antipyretic properties.

Foundation for Revitalization of Local Health Traditions (FRLHT) Banglore in collaboration with HFRI, Shimla organized a Conservation Assessment and Management Prioritization (CAMP) Workshop for medicinal plants of North Western Himalayan States of Jammu and Kashmir, Himachal Pradesh and Uttarakhand during 22–25 May, 2003. I was one of the invitees to the workshop which deliberated upon the IUCN red listing criteria and categories as per IUCN version 3.1 of 2000 and assigned various threatened category status of well known medicinal plants of North-Western Himalaya. International Union for Conservation of Nature (IUCN) Red list categories are intended to classify species at high risk of global extinction. The red list categories may focus attention and management on taxa experiencing highest risk of extinction but it is not the sole means of setting priorities for conservation measures for their protection. Since, their adoption by IUCN Council in 1994, the IUCN Red list categories have become widely recognized internationally. Various medicinal plants assessed for their degree of extinction threat fall under one of the following categories:

Extinct (EX)

A species is extinct when there is sufficient evidence that the last individual has died down and exhaustive surveys have failed to record its existence *e.g. Dicliptera abuensis* (*Dicliptera bupleuroides* is present in Western Himalayas).

Extinct in the Wild (EW)

A species is extinct in the wild when it is known only to survive under cultivation or captivity and exhaustive surveys have failed to locate it in the wild *e.g. Ginkgo biloba* (oldest *Ginkgo biloba* tree is seen at Lal Mandi, Srinagar).

Critically Endangered (CR)

A species is said to be critically endangered when the best available evidence suggests that its population reduction is ≥ 90 per cent over the last 10 years or three generations whichever is longer, in its geographic range, its extent of occurrence is < 100 km^2 and area of occupancy is < 10 km^2 and habitats are severely fragmented *e.g. Arnebia benthami* (Vern. Ratanjot, Gahozawain).

Endangered (EN)

A species is said to be in endangered category if best available evidence suggests that it is facing a very high risk of extinction in the wild. Determining parameters are population reduction ≥ 70 per cent over the last 10 years or three generations and extent of geographic occurrence is less than 5000 km^2 and area of occupancy is less than 500 km^2. Habitat Fragmentation is also taken into consideration *e.g. Atropa acuminata* (Vern. Bisral, Tookan sool and Belladona).

Vulnerable (V)

A species is said to be vulnerable if all evidence indicates suspected population reduction of ≥ 50 per cent over the last 10 years or 3 generations, extent of occurrence less than 20000 km^2 and area of occupancy less than 2000 km^2 *e.g. Rheum webbianum* (Vern. Churko, Pambhaak).

Near Threatened (NT)

A species is considered near threatened if it has been evaluated against all the determining criteria but does not qualify for critically endangered, Endangered or Vulnerable category at present *e.g. Rhododendron lepidotum* (Vern. Nichhni).

Least Concern (LC)

A species falls in the least concern category when after evaluation, it does not qualify for any other category enumerated above. Widespread and abundant taxa are included in this category *e.g. Eremurus himalaicus* (Vern. Kahloon, Bulkut and Chhilhaak).

Data Deficient (DD)

A species is data deficient when there is inadequate data available to determine its IUCN category. It is therefore, not a category of threat and listing species as data deficient indicate that more information and research is required to determine its threatened status in future *e.g. Rheum spiciforme* (Vern. Churko, Pambhaak).

Prominent medicinal plants of North Western Himalayas which were evaluated in CAMP workshop at Shimla are enumerated below:

Critically Endangered Medicinal Plants

***Aconitum chasmanthum* stapf ex Holmes (Family: Ranunculaceae) Vern. Methatelia, Vatsnabh.**

High altitude alpine herb with irregular, spurless and blue flowers. Found in Gilgit (PoK), Gurez, Sonamarg, Baltal in Kashmir and Zanskar in Ladakh. Roots medicinal. Its population is on the decline.

***Aconitum heterophyllum* Wall ex Royle (Family: Ranunculaceae) Vern. Patis, Atees, Buma.**

High altitude herb found in open meadows as well as in wooded forest areas. It is globally critically endangered but is endangered in western Himalayas. Roots medicinal.

***Arnebia benthami* (Wall ex G. Don) Johns (Family: Boraginaceae) Vern. Ratanjot, Ghaajwan, Balchhar.**

High altitude herb of open rocky slopes. Flowering spikes woolly and shaggy and used in the treatment of cardiac ailments. It is found in Wadwan, Sonamarg, Kaobal (Tilal), Zanskar (Kargil), Padder, Kishtwar, Duksum, (Kashmir), Harkidoon, Gangotri, Badrinath (Uttarakhand). In HP, it is reported from Lahul Spiti, Kinnaur, Chamba and upper Parvati Valley.

Arnebia euchroma **(Royle) John (Family: Boraginaceae)**
Vern. Ratmundi, Ratanjot

Cold desert plant found in open rocky mountains. Seen near Tsomorari lake in Ladakh. Also found in Lahul-spiti, Pin valley and Satrundi area of H.P. It is reported from Garhwal and Kumaon in the Uttarakhand state.

Atropa acuminata **Royle ex Lindley (Family: Solanaceae)**
Vern. Jalakaffal, Bisral.

Tall perennial of great medicinal value. It is found in Poonch, Gulmarg, Gurez, Machhal, Marwah, Kukkamag, Chhattargala, Parkachik (J&K), Pangi, Sangla and Narkanda (HP). Favoured habitat is along margins of Oak and Kail forests. Not reported from Uttarakhand in the wild but stands introduced in the germplasm gardens (G.S. Rawat). It is critically endangered only in Himachal Pradesh and falls in simple endangered category in J&K.

Dactylorhiza hatagirea **(D. Don) Soo (Family: Orchidaceae)**
Vern. Salampanja, Hathpanja

Handsome orchid of temperate moist meadows in the Himalaya. It is found from Pakistan to Nepal and China. Pink red flowering spikes and palm like roots are diagnostic features. It is found in Nubra, Sankoo (Ladakh), Gurez, Khilanmarg, Sonamarg, Pahalgam (Kashmir), Sangla valley, Baspa valley, Manali, Katgaon (H.P.), Valley of flowers, Dayara, Bhuj Vasa (Uttarakhand).

Gentiana kurroo **Royle (Family: Gentianaceae)**
Vern. Karu, Trayaman, Neelkanth

One of the most beautiful Gentian of the North Western Himalaya. Loss of habitat due to fragmentation and overexploitation for ayurvedic drugs has reduced its population drastically (≥ 80 per cent) in J&K, H.P. and Uttarakhand states. It was once common in Pancheri and Tarmain Pine forests in the 1980's but now very low populations are available in these areas. It is found in Chenab valley. Pir Panchal and Kashmir valley of J&K, Kalka Pattan, Chanshal, Karol and Chitkul areas of H.P. and Garwhal, Uttarkashi, Gangotri and Pindari area of Uttarakhand. It is globally threatened (FRLHT, 2003).

Lilium polyphyllum **D.Don ex Royle (Family: Liliaceae)**
Vern. Gioter, Kakoli

Tall bulbous perennial with fragrant drooping purple spotted white flowers. It has been reported from Chhatru, Simthon Pass and Pir-Panchal area of J&K, Chakrata, Deoban and Pauri Garwhal area of Uttarakhand and Dhauladhar and Shimla forests of H.P. Extensive field surveys carried out by the author in Simthon Pass, Machhal (Kashmir) and Sarthal, Patnitop (Jammu) failed to locate this species which suggests it has become very rare in the J&K forests. It is treated as critically endangered species in the North-Western Himalaya.

***Saussurea costus* (Falc.) Lipsch (Family: Asteraceae)**
Vern. Kuth, Minael, Rusta

Robust perennial of moist temperate forests, whose roots are highly aromatic and are in great demand in the' world market. Once, this species (vern. Kuth, Minael, Rusta) used to earn major foreign exchange for the state of J&K. Kuth fields were once established in Gurez and Astore areas of J&K and Kuth Act was introduced to regulate its commerce. One of the constituent of Asthavarga, it is used in the preparation of Chyavanprash. In recent surveys undertaken by State Forest Research Institute, J&K, its rich existence in the Gurez-Tilel area of Bandipore Forest Division stands established. Elsewhere its population has declined. Illicit wild root collections is the major threat to its survival. Its cultivation need to be revived in Kashmir. It is reported to exist in Drass, Zanskar and Sum valley (Kargil), Seoj, Sarthal (Jammu), Kalohoi mountains (Kashmir) in J&K. In H.P. this species is found to exist in Lahul Spiti, Kinnaur and Chamba, but it has not been recorded from Uttarakhand forests. Major wild germplasm is seen only in J&K.

***Swertia chirayita* (Roxb. Ex Fleming) Karsten (Family: Gentianaceae)**
Vern. Kirat, Chirata, Bhunimba, Tikta

Most important and widely used medicinal herb of North-Western Himalaya. Loss of habitat and overexploitation for medicine has decimated its wild populations and now it is very rare to collect this species in the wild. This species exhibits 4-petalled flowers and roots are antipyretic. In J&K its existence is doubtful but it is found in Kullu area of H.P. In Uttarakhand, it stands reported from Chakrata hills, Dodital, Valley of flowers and Kedar Valley.

Endangered Medicinal Plants

***Aconitum deinorrhizum* Stapf (Family: Ranunculaceae)**
Vern. Mohra, Patees

Alpine, sub-alpine herb endemic to Bashar Himalaya. It is reported to exist in Bhaderwah and Pir-Panchal area of J&K and Shangli thatch, Chenshol and Tangaankhai area of H.P. It is not found in Uttarakhand state. Severely fragmented habitat is major concern. It stands included in negative list of Export, 1998. Its cultivation / propagation is very difficult as compared to *Aconitum heterophyllum* (Patees). Microhabitat analysis may unravel the mystery in the future.

***Angelica glauca* Edgew (Family: Apiaceae)**
Vern. Choru, Chohre

Tall branched aromatic perennial of humus rich forests. Roots are heavily exploited for aromatic oil and medicinal use. Both *in situ* and *ex situ* conservation is practiced in North-Western Himalayan states. In J&K, it is found in Dahi nala, Dawar, Chorwan (Gurez), Angaikot, Raman nala (Tilel). In H.P. its populations stands located at Chamba and Parvati valley. Garhwal and Kumaun is also known for its wild occurrence.

***Betula utilis* D. Don (Family; Betulaceae)**
Vern. Burj, Joj

High altitude tree of Himalayas which is reported in Afghanistan and Central Asia outside India. It is found mostly along Timberline but also found in pure and mixed coniferous forests (*e.g.* on way to Pahalgam). Usual place of occurrence in Sonamarg, Khilanmarg, Dawar, Tilel, Suru, Nubra and Valley of Flowers, Bhujvasa and Dayara in Uttarakhand. Climate change, snow damage, firewood extraction are major threats for its survival in the wilderness. Trade in Amritsar market has increased from 30 quintals in 1996 to 60 quintals in 2003 and rate per kg bark had increased from Rs 30 to Rs 90 in the same period (HFRI and FRLHT studies).

***Bunium persicum* (Bioss.) Fedts. (Family: Apiacaae)**
Vern. Kala Zira.

Bulbous perennial herb of dry temperate areas which is known for its medicinal and spice value. Its presence in Paddar area of Jammu region Gurez area of Kashmir is well known. In H.P., it is found in Shong valley, Lahual, Kinnaur and Pangi valley.

Cinnamomum tamala **(Buch-Ham.) Nees (Family: Lauraceae)**
Vern. Tejpattar, Tejpat

Aromatic tree of moist shady ravines in the temperate area. Its occurrence in Bambro, Balhota (Ramnagar), Banjal gala (Basohli), Maungri Basnot (Udhampur) stand documented. It is also found in Pabbar Valley, Jogindernagar and Chamba forests of H.P. It grows in Uttarakhand forests as well. Its occurrence is highly sporadic in our forests. Its leaves and bark are traded for food and medicine. It is adultrated with non-aromatic leaves of *Cocculus laurifolius* (Vern. Chopra) which grows in sub-tropical areas of all the three Himalayan states. Its seedlings recruitment is moderate to good but establishment is very poor. 100–200 metric tonnes material is traded annually in our area.

Datisca cannabina **L. (Family: Datiscaceae)**
Vern. Kauri Karer, Bhangjala

Tall herb which is found colonized in freshly eroded hill slopes and along stream courses. Its pollination is a main limiting factor for its successful reproductive behaviour. It is dioecious and monotypic plant showing fragmented habitat distribution in places like Makarkote (Ramban), Koragbal (Gurez), Ghratnar (Tilel), Kalaroos (Kupwara) in J&K, Kullu, Kinnaur, Lahaul spiti, Pattan valley in H. P. and Hanuman Chatti, Karna Prayag, Binsar, Gadolia, Garhwal-Kumaon in Uttarakhand.

Dioscorea deltoidea **Wall ex Griseli (Family: Dioscoreaceae)**
Vern. Krits, Shingli mingle

It is one of the most important medicinal climber of temperate forests whether mixed broadleaved or coniferous. Its underground tuber is a source of steroid hormones which are used in the preparation of contraceptive pills. It is heavily exploited for its tubers which are used in washing woolens and marketed for the drug manufacture. Its favourable habitats in J&K state are Bhaderwah, Padder, Sarthal, Sannasar, Dudu (Jammu region), Dachigam, Toshmaidan, Dudhpathri, Naranar, Pahalgam, Machhal (Kashmir region), Dawar, Chorwan, Dahi nala (Gurez). Major habitat in H.P. is found at places like Shimla, Chamba, Dalhousie, Mandi, Kangra. In Uttrakhand, Dioscorea is seen at places like Tehri, Pauri Garhwal, Pithoragarh and Nainital. Its propagation protocol is available and is cultivated in the forest nurseries.

***Ephedra gerardiana* Wall ex Stapf (Family: Ephedraceae)**
Vern. Chhepath, Somlata, Chhi, Sutchur

Cold desert plant of gymnospermic nature found in southern exposed slopes amongst rocks and dry sandy situations at places like Lamayuru, Nubra, Zanskar (Ladakh), Akpa, Pooh, Khab, Tabo, Roopa (H.P.) and Garhwal Kumaon and Chamoli (Uttarakhand). Entire plant is traded for medicinal use. It is distributed all along Afghanistan, Nepal, Bhutan to China. Propagation is moderately difficult. *Ephedra intermedia* Schrenk et Mey is used as an adulterant.

***Fritillaria roylei* Hook. (Family: Liliaceae)**
Vern. Kakoli, Sheethkar

Sub-alpine, alpine perennial of rocky habitats with whorled leaves and drooping flowers. In J&K prominent places of its occurrence are Poonchh, Seoj dhar, Kailash (Jammu), Simthan pass, Kalohai mountains, Rajdhan pass (Kashmir), Pokhri-pah, Kaobal (Tilel) and Dawar Gurez forests. In H.P. its occurrence is reported from Chhajpur, Gulaba (Kullu) and Churdhar area. Its distribution in Uttrakhand is seen at valley of flowers, Tungnath, Gangotri and Kush Kalyan. Its bulbs are medicinal and are used in the preparation of the Ayurvedic tonic, Chyavanprash. Propagation of this species in nurseries is very difficult.

***Habenaria intermedia* D. Don (Family: Orchidaceae)**
Vern. Devsund, Riddhi

Another medicinal orchid after Palm orchid (*Dactylorhiza hatagirea* vern. Salmpanja) which is found distributed in moist meadows of temperate. Its tuberous roots (Riddhi, Devsund, Padma) are traded in the market. Being orchid, it is very sensitive to minor habitat changes. Once common in Marta village (Ramnagar), it is vanishing very fast from our landscape. Its existence in Basohli and Billawar is from places like Ram Rachna and Dhaggar. In H.P. it is seen in places like Chamba, Dalhousie, Kinnaur and Shimla whereas in Uttarakhand it is found growing in few locations in Garhwal and Kumaon. It stands included in the National negative list of exports.

***Juniperus polycarpos* C. Koch (Family: Cupressaceae)**
Vern. Shukpa, Padam, Bethru

Medicinal gymnospermic tree of the dry cold area like Foker, Zanskar, Drass, Batalik (Kargil), Dawar (Gurez) in J&K; Lahul Spiti,

Pangi, Kinnaur, Nako, Darchha in H.P. and Nanda Devi, Nilang, Niti and upper Johar area in Uttarakhand. Its bark, barriers and leaves are traded for essential oil and medicinal use. It is also harvested for timber. Lopping and debarking results in ultimate drying of natural trees. It shows very poor seed germination due to embryo and seed dormancy. Nursery Technology needs to be standardized. Existing cultivation is not known.

Jurinea dolomiaea **Boiss (Family: Asteraceae)**
Vern. Guggal Dhoop, Dhoop

Useful perennial herb of grassy alpine slopes which is exploited heavily for medicine and Dhoop agarbatti products. Its roots are aromatic and used as an incense. Favourable habitats of its occurrence are at Peer Panchal, Wadwan, Sarthal, Saamnabanj, Seoj, Neel, Margam top, Simthan pass, Rajdhan in J&K, Pangi, Kangra and Parvati valley in H.P. and Deyara and Tungnath in Uttarakhand.

Litsea glutinosa **(Lour) Robinson (Family: Lauraceae)**
Vern. Medda Sakk, Maida, Lakkri, Rehan, Chirar

Aromatic tree of lower hills whose bark is medicinal and used in the treatment of bone fracture by the natives. Inner bark is also used in perfumery industry. Poor regeneration has led to its drastic population decline in Shiwaliks. It is found in places like Raang in Ramnagar, Kalakote (Rajouri), Chew (Billawar) in J&K; some locations of H.P. and Kumaon and Gharwal hills of Uttarakhand. It can easily be propagated and need to introduce in our forest nurseries.

Meconopsis acuteata **Royle (Family: Papaveraceae)**
Vern. Gulneelmi, Swarnkanta

One of the most magnificent blue poppy of the Himalayas which grows at higher altitudes in rocky regions and moraine margins. It is seen in places like Simthan Pass, Madwa–Dachhan, Chandanwari (Kashmir) Zanskar, Parkachik in J&K, Kullu, Kinnaur, Lahul-Spiti, Rohtang pass, Mani Mahesh, Sangla, Choordhar in H. P. and Valley of flowers in Uttarakhand. Its roots yield Xanthotoxin. Another plant *M. latifolia* is endemic to J&K and is seen in areas like Gangbal (O.P. Sharma 2007).

***Nardostachys grandiflora* (D. Don) Dc (Family: Valerianceae)**
Vern. Jatamasi, Nehani, Balchhar

Erect herb with thick spindle shaped tuberous root exploited for the preparation of Ayurvedic drugs. It figures in Appendix–II of CITES and in the Negative list of exports, 1998. It is not found in the forests of J&K but grows in the temperate forest areas of H.P. and Uttarakhand.

***Paris polyphylla* Smith (Family: Liliaceae)**
Vern. Nagpattar, Meethi Bach, Nagchhatri

Small bulbous herb with whorled leaves. It is found in humus rich moist shady temperate forests at places like Poonch, Bhaderwah, Pir Panchal (Jammu), Kullu, Chamba Shimla (H.P.), Uttarakashi, Valley of Flowers, Dodital and Tungnath (Uttarakhand).

***Podophyllum hexandrum* Royle (Family: Podophyllaceae)**
Vern. Bankakri, Gulkad

High altitude herb of moist shady places with humus rich soil. Its rhizome and roots are a source of anticancer drug. Its propagation is not easy and need lot of management interventions. It is being introduced in Vanaspati vans. Loss of habitat and overexploitation has pushed this plant into an endangered category. It is found in Gurez-Tilet, Khlanmarg, Gulmarg, Phalgam, Aru in Kashmir and Sarthal, Bhaderwah, Jai, Seojdhar in Jammu region. It is found growing in Pangi, Chamba, Kinnaur, Kullu, Churdhar in HP and Kedarnath, Valley of Flowers, Harkidoon, Kauripass and Dodital in Uttarakhand.

***Rheum emodi* Wall ex meissn. (Family: Polygonaceae)**
Vern. Chukri, Churko, Pambehalan

Robust alpine, sub-alpine herb of steep dry mountain slopes. Loss of habitat, grazing pressure and overexploitation for food and medicine has deceased its populations in the wild. Roots are traded as Revandchini and tender foliage and stems are eaten as Pambhaak or Churko by the natives. Rhizomes are medicinal. It is found in Sarthal, Padder, Bhaderwah, Gurez and Tilel area of J&K Pabbar valley, Pin valley and Parvati valley in HP and Kedar valley and Valley of Flowers in Uttarakhand.

***Taxus wallichiana* Zucc. (Family: Taxaceae)**
Vern. Birmi, Rakhal, Thuneer

Another medicinal gymnosperm of temperate forests which grows as under-storey tree in Fir/Spruce and Kharsu oak forests. It is very slow growing tree and is prone to the attack of insect and pathogens. Techniques for vegetative propagation stand standardized and are being followed in forest nurseries. Its bark earlier used as tea substitute (Barmi Cha) has now turned out a source of anticancer drug Taxol. It is found growing in Pancheri, Lander, Sannasar, Tangmarg, Gulmarg, Kangan, Gurez in J&K and Chhajpur, Kanasar, Kullu in H. P. and Gharwal and Kumaon in Uttarakhand.

Vulnerable Medicinal Plants

Many medicinal plants like *Aconitum violaceum* (Mohand, Mohri), *Allium stracheyi* (Kiyar, Pharrnu), *Artemisia maritima* (Seski, Moori), *Bergenia stracheyi* (Sapdotri, Katkotar), *Ceropegia bulbosa* (Galot bel), *Colchicum luteum* (Suranjan-e-talakh), *Embelia tsjeriam-cottam* (Baibiring), *Eremostachys superba* (Gaajarmoola), *Ferula jaeshkeana* (Sampheru), *Gloriosa superba* (Kukkarsira, Langli), *Heracleum lanatum* (Patrala), *Hippophae rhamnoides* (Seabuckthon, Leh-berry), *Physochlaina praelta* (Langtang), *Polygonatum verticillatum* (Mirkuli Salam-mishri), *Rauvolfia serpentina* (Sarpgandha), *Rheum spiciforme* (Churko), *Rhodiola heterodonta* (Shrolu), *Rhododendron campanulatum* (Sarangar), *Roylea cinerea* (Titpati), *Saussurea obvallata* (Brahmkamal), *Valeriana jatamansi* (Mushakbala), *Zanthoxylum armatum* (Tinbru) are dwindling in the wild as a result of habitat loss, loss of pollinators, over exploitation for trade and commerce and overall shrinking forest areas.

Conservation Concerns

Going by the above account, it seems ample clear that our forest habitats are shrinking day by day, wild germplasm of medicinal plants is being exploited ruthlessly, thereby pushing many species to the brink of extinction. Some of the reputed medicinal plants have already experienced drastic decline in their population status and have become critically endangered. Majority of our medicinal plants have become endangered and need immediate conservation action at various levels. To address the need for conservation of germplasm of native medicinal plants, it is imperative to have a network of *in situ* conservation sites or field gene banks in different agro climatic zones in the Western Himalaya. Depending on the size and diversity of available habitats occurring in a state, there is a need to establish around 10 or more field gene banks in the State of J&K, H. P. and Uttarakhand. The gene banks for plant species serve the objective of *ex situ* conservation of medicinal plants but these banks are quite expensive to create and maintain. A network of *in situ* (field) gene banks in forest habitats is the most cost effective way to manage the intraspecific and inter-specific plant diversity. The *in situ* gene banks sites can also be used as study sites to understand the reproductive biology of the species which are of high conservation value (*e.g.* Critically Endangered Medicinal Plants) and can guide strategies towards their recovery and long term conservation. A comprehensive documentation of floristics of such selected sites alongwith rapid threat assessment exercises to assess the risk of extinction of prioritized species will help in guiding focused conservation action programs to save the wild medicinal plants of Western Himalaya.

National Medicinal Plant Board setup under a Government resolution notified on 24th November 2000 is to be viewed in the backdrop of conservation concern of our medicinal flora. It has prioritized 32 medicinal plants for providing financial assistance to promote integrated development by co-ordaining, stimulating production, processing, marketing and establishing a sound infrastructure of the medicinal plant sector in the country. The board aims to make the cultivation of medicinal plants and its sustainble management a people's movement. The list of prioritized medicinal plants is as under:

(*i*) Amla–*Emblica officinalis* Gaetrn.
(*ii*) Ashok–*Saraca asoca* (Roxb.) De Wilde.
(*iii*) Ashwagandha–*Withania somnifera* (Linn.) Dunal
(*iv*) Atees–*Aconitum heterophy/lum* Wall ex Royle
(*v*) Bael–*Aegle marmelos* (Linn.) Corr.
(*vi*) Bhumi amalki–*Phyllanthus amarus* Shum and Thonn.
(*vii*) Brahmi–*Bacopa monnieri* (L.) Pennell
(*viii*) Chandan–*Santalum album* Linn.
(*ix*) Chirata–*Swertia Chirayata* Buch Ham.
(*x*) Giloe–*Tinospora cordifolia* Miers

(*xi*) Gudmar–*Gymnema sylvestre* R. Br.
(*xii*) Guggal–*Commiphora wightii* (Am.) Bhandari
(*xiii*) Isabgol–*Plantago ovata* Forsk.
(*xiv*) Jatamansi–*Nardostachys jatamansi* DC.
(*xv*) Kalihari–*Gloriosa superba* Linn.
(*xvi*) Kalmegh–*Andrographis paniculata* Wall ex Nees
(*xvii*) Kokum–*Garcinia indica* Chois
(*xviii*) Kuth–*Saussurea costus* C. B. Clarke
(*xix*) Kutki–*Picrorhiza kurroa* Benth ex Royle
(*xx*) Makoy–*Solanum nigrum* Linn.
(*xxi*) Mulethi–*Glycyrrhiza glabra* Linn.
(*xxii*) Musli Safed–*Chlorophytum borivillianum* Sant
(*xxiii*) Patharchur–*Coleus barbatus* Benth.
(*xxiv*) Pippali–*Piper longum* Linn.
(*xxv*) Rasaunt–*Berberis aristata* Dc.
(*xxvi*) Saffron–*Crocus sativus* Linn.
(*xxvii*) Sarpgandha–*Rauvolfia serpentina* Benth. Ex Kurz.
(*xxviii*) Senna–*Cassia angustifolia* Vehl.
(*xxix*) Shatavari–*Asparagus racemosus* Willd.
(*xxx*) Tulsi–*Ocimum sanctum* Linn.
(*xxxi*) Vai Vidang–*Embelia ribes* Burro. F.
(*xxxii*) Vatsnabh–*Aconitum ferox* Wall.

In J&K state Medicinal Plant Board was established in January, 2001 and is extending financial help to the Entrepreneurs, NGO's, farmers, Research Institutions, Forest agencies to undertake *in situ* and *ex situ* conservation of medicinal plants. *In situ* conservation sites stand identified at Ramwari, Shajmarg, Son-Rasin (Sind), Khodmarg, Wangat, Badpathri (Manasbal). *Ex situ* conservation sites are established at places like Urpash, Ganangeer (Manasbal), Nilgar Baltal, Gund (Sind). Hi-tech nurseries stand established at Ganderbal and Baltal (JKSMPB, Annual Report, 2006). Village Van Committees to implement the program stand constituted at places like Ganangir (Mohd. Ibrahim Sheikh), Wangat (Gulam Qadir Mir), Sonamarg

(Gulam Nabi Jabbar), Ramwari Gund (Mohd. Sadiq Kaloo), Kulan (Moulvi A.H. Sheikh), Urpash (Ms. Roshan Ara) and Naranag (Abdul Qayoom Jagal). National Seminars and Workshops are conducted to create desired impact on the growth and development of medicinal plant sector.

Regional Research Laboratory now named as Indian Institute of Integrative Medicine, Jammu is doing useful work on various endangered medicinal plants. Tissue culture techniques and propagation modules of some selected medicinal plants like Chirata, Kutki are being standardized. Herbal Nurseries stand established at various places in J&K.

In the state of HP and Uttarakhand, various initiatives are being taken to conserve the Himalayan medicinal flora on similar lines. People's participation and Government support in the form of financial and technical know-how are becoming central to all programs in the state.

References

Ahmedullah, M., 1997. *Biodiversity of Jammu and Kashmir: A Profile.*

Chaurasia, Om Parkash, 1996. *Cold Desert Plants, Vols. I-V.* Field Research, FRL, DRDO, Leh.

Jain, S.K., 1994. *Medicinal Plants.* National Book Trust, New Delhi.

Munshi, A.H. and Javed, G.N., 1986. *Systematic Studies in Polygonaceaae of Kashmir Himalaya.* Scientific Publishers, Jodhpur.

Panwar, Vijendrer Paul *et al.*, 2007. *Aushdiya Pandhon ki Krishikaran Takneek.* HFRI, Conifer Campus, Panthaghati, Shimla.

Ved, D.K. *et al.*, 2003. *A Synthesis of Regional Expertise in Medicinal Plants: Taxonomy and Distribution Through a CAMP Workshop,* held at Shimla, 22–25. May.

Vidyarthi, O.P., 2003. *Medicinal Plants of Plains and Hills.*

Chapter 10

Rich and Varied Fish Fauna of Himalayas

B.L. Kaul

ABSTRACT

The Fish fauna of North Western Himalaya which includes Jammu and Kashmir State is of special interest from biogeographical, geological and evolutionary points of view. The water bodies both lentic and lotic are increasingly becoming polluted due to siltation, inflow of nutrients from hill side, landslides and increasing human activities such as agriculture, deforestation and tourism etc. This is posing threat to the biogeographically significant fish fauna and calls for immediate remedial measures.

Introduction

The territories of Jammu, Kashmir and Ladakh a mountainous area of 2, 22, 236 sq.km. (Including parts occupied by Pakistan and China) in North Western Himalaya occupies an almost central position in the continent of Asia. Dotted with low lying valleys here and there this irregular area lies between 32°17′ N and 37°6′ latitudes and 73°26′ E and 80°30′ E longitudes. It is the northernmost state of India and is strategically the most important. Here the national boundaries touch the neighbouring countries of Tibet in the east, China and Afghanistan in the north and Pakistan in the west. To its south lies Punjab, Himachal Pradesh and other states of the country. This locational aspect of Jammu and Kashmir state accounts for its geopolitical and biogeographical significance. The fauna, especially Ichthyofauna is a mixture of palacarctic, endemic and oriental elements. It makes an interesting study.

Geography of Kashmir Himalaya

Geographically, the Himalaya is divided into: 1. The Eastern or the Assam Himalaya. 2. The Central or The Nepal Himalaya. 3. The Kumaon or The Western Himalaya and 4. The north-western or Punjab Himalaya which includes Jammu and Kashmir and Himachal Pradesh. The N.W. Himalaya is the division west of the defile of the river Sutlej west of which the river Indus curves round south west ward. In the state of Jammu and Kashmir there are two mountains masses *viz.* the *Korakoram* and the Himalaya, with the river Indus in between. On the southern flanks of the main Himalaya lies the valley of Kashmir proper and is walled in the Pir Panjal range. The Siwalik zone also known as the Himalayan foothills stretches from the Ravi to Jhelum. The Pir Panjal bifurcates from the main Himalaya but structurally and lithologically, it is most complex. The rivers Chenab and Jhelum drain these mountain ranges. The main Himalaya running between the peaks of Nun-Kun (7135 m) and Nanga Parbhat (8127 m) bound Ladakh region, in the south. Branching from it and later on running parallel to it is the *Zanskar range.* A number of mountain chains, including the Himalayas radiate from the Pamir Knot in various directions. The Himalayas are reinforced on the north by a mountain chain, known as the *Korakoram Range.*

Rivers and Lakes

The chief rivers of Jammu and Kashmir state are the Indus, the Jhelum and the Chenab all of which tread their way though the territories of the state and form a part of the Indus drainage system. The Indus is a great river of the north western India rising in Tibet and passing through Ladakh division of the state. It rises from the mountain district around *Mansarover* wherefrom the Sutlej and the Brahmputra also arise. The Jhelum has its source at a spring called Verinag in the Pir Panjal range. The river flows in loops over a distance of 176 kms, across the valley of Kashmir through the Wular Lake. The Chenab rises in the Himalayan contour of Lahol in two streams, the Chandra and Baga. They flow more or less parallel to each other and ultimately meet at Tandi to form the Chandrabaga or Chenab which flows through the Pangi Valley in the Chamba district of Himachal Pradesh before entering Padar tehsil of Kashmir. Besides these rivers, there are numerous streams forming their tributaries.

The state of Jammu and Kashmir has many lakes which includes the *Wular* which is the largest fresh water lake in India. The beautiful *Dal Lake* which is shrinking due to pollution. The other Lakes are the *Anchar*, the *Mansbal*, the *Kaunsernag*, the *Sheshnag*, and *Gangabal* in Kashmir, the *Mansar* and *Sanasar* in Jammu and famous salt water lake *Pang Pong* in Ladakh.

Ichthyofauna

The varied and difficult terrain of Jammu and Kashmir in North Western Himalaya sustains a rich Ichthyofauna which is of great interest biogeographically. While the fish fauna of the plains of Jammu shows a close affinity with the fish fauna of the neighboring Punjab, the fish fauna of Kashmir, Ladakh and hill districts of Jammu are a mixture of *palaearctic*, *endemic* and *oriental* elements. Similarly fish fauna in Himachal Pradesh, Uttarakhand, Nepal, Sikkim, Bhutan and hill districts of West Bengal. Of the 35 or so species of Kashmir fishes the ratio of species between (1) Central Asiatic, (2) Oriental and (3) endemic as established by Hora (1937) was 9 : 7 : 1 = 17. Das (1963) found the ratio to be 22:8:5=35. Thus the Central Asiatic or Palaearctic element (Das 1960) 22 is the highest representation in Kashmir fishes in which *Schizothoracinae* are predominant. Das (1970) has put forth the *Parmir Kashmir Theory* for the evolution of Kashmir fishes and has taken into consideration physical, geographical and ecological factors to support his theory.

According to Jayaram (1974) the palaearctic fish element in the Himalayan fish fauna are:

1. *Schizopygopsis*.
2. *Schizothorax*
3. *Ptychobarbus*
4. *Cyprinion*

Schizopygopsis and *Ptychobarbus* are endemic in the Turkmanian subregion, *Cyprinion* is a Mediterranean genus of Mediterrarean subregion. Of the eighteen species of *Schizothorax* known, ten are found in the Mediterrarian subregion. All the seven species of

Cyprinion, so far known occurs in the northwest border land and extend to sourtheast Iran. Three species of *Ptychobarbus* and solitary species of *Schizopygopsis* are endemic in the Northwest Himalaya, mainly in the river Yarkand and other rivers in the area. These genera have not so far been recorded from areas east of the great defile of the river Sutlej. They also occur at lower elevations only very rarely.

Schizothorax and *Ptychobarbus* are generally found in lakes and in large rivers, with backwaters.

Introduction of the Chinese carp *Cyprinus carpio* in 1955 in the Kashmir valley has forced *Schizothorax* to vacate its habitat and shift to cold water streams.

Schizopygopsis is equally at home in lakes, large rivers and even in small rapids rich in algal growth. The various steams and lakes in the northwest Himalayas, where this fish occurs, are not very widely separated from each other by high ridges. Many species are, therefore, common to different streams and unlike species in the plains, they are not geographically restricted to specified water sheds. It would seem that the lakes and rivers of the N.W. Himalaya are relics of a former larger and continuous mass of water (Hora and Mukherji 1935)

The following genera are common to the Palaearctic area and Oriental of India. Some of them are also Ethiopian.

Family Cyprinidae

1. *Garra*
2. *Crossocheilus*
3. *Oreinus*
4. *Schizothorichthys*
5. *Diptychus*
6. *Labeo*
7. *Catla*
8. *Amblypharynygodon*
9. *Tor*
10. *Esomus*
11. *Rasbora*
12. *Barilius*

13. *Aspidoparia*
14. *Danio*
15. *Chela*
16. *Oxygaster*

Family Cobitidae

1. *Botia*
2. *Lepidocephalichthys*
3. *Naemacheilus*

Family Bagridae

1. *Rita*
2. *Mystus*

Family Sisoridae

1. *Bagarius*
2. *Glyptothorax*
3. *Glyptosternum*

Family Clariidae

1. *Clarius*

Family Channidae

1. *Channa*

The genera *Oreinus, Glyptosternum, Glypothorax, Schizothoricthys, Botia, Nemacheilus* and *Diptychus* occur in a more dynamic environment than the four genera that are restricted to the Palaearctic areas of India. *Diptychus* is a monotypic genus that occurs in rapid streams (Hora and Mukherji 1935). *Glyptothorax kashmirensis* inhabits torrential streams in the Kashmir valley. *Glyptosternum* is a palaearctic genus, apparently derived from the Indo-Chinese *Glyptothorax*-like ancestors. Of the three species known at present *Glytosternum akhtari* is confined to the river Oxus in Afghanistan, *Glyptosternum reticulatum* is found in Afghanistan, *Glytosternum reticultum* is found in Ladakh region of Kashmir, Chitral and Afghanistan and *G. maculatum* is the only species, which extends to Sikkim in the Eastern Himalayas, far from its main home in Tibet. *Nemacheilus* is found in the main in central Asia, in Trans-caucasia and also in the Far East. Although it prefers torrential streams yet a few species have become adapted to lentic waters in lakes and reservoirs. *Botia* is also found in swift flowing waters, over stony and sandy bottoms. In the Kashmir valley it has turned a resident of calm waters of lakes but migrates to river Jhelum for breeding (Kaul 1972).

Of the three species of *Oreinus* so far known, one is endemic in the Mediterranean subregion, one in the Indo-Chinese sub-region and only *O. plagiostomus* extends to the eastern Himalaya from the west, where it also occurs as far as Arabia and eastern Iran. *Oreinus plagiostamus* has diverged into many local forms in the N.W Himalaya and hybridization is also not uncommon (Jayaram and Majumder, 1964). *Schizothoricthys* is a derivative from the *Palaearctic Schizothorax* and only one species extends into eastern Himalaya. *Diptychus maculatus* has also similarly extended into the eastern Himalaya.

The fish like *Rita* and *Mystus* are ancient but well stabilized genera and represent elements of ethiopian origin. *Chrysichthys* from Africa is considered as the ancestor of *Rita* and the genus *Mystus* is related similarly to the African *Porous* (Jayaram 1966). Thirty-three species of *Clarias* are known from Africa (Buolenger 1911) but only seven are found within the limits of Indian subcontinent. *Clarias batrachus* is found continuously from Sind to Malay Archipelago and from Kashmir to Srilanka. *Garra, Labeo, Puntius, Barilius* are also similarly widely distributed in Africa, Iran and Northwest India. *Nemacheilus* extends on the other hand, from the fresh waters of Europe to Asia.

Threat to Fish Fauna of Himalayas

The Ichthyofauna of the North western Himalaya which includes Jammu and Kashmir State and Himachal Pradesh is of special interest from biographical, geological and evolutionary point of view. The story of the evolution of fish fauna in the region is closely linked with the evolution of the Himalayas and the geological changes that have taken place there. This is amply proved by the palaeontological and faunal studies made by various workers from time to time. The presence of palaearctic elements in the fish fauna of the region also lends credence to this fact. It is rather unfortunate that the Inchthyofauna of Himalayas is facing threat due to increasing pollution of their habitat. Ecodegradation of the Himalayas is affecting the quality of both lentic and lotic water bodies which in turn is threatening the biogeographically significant fish fauna of the region. The water bodies are increasingly becoming polluted due to over population, siltation, deforestation, landslides, agricultural activities, inflow of nutrients from hillsides, road building and tourism. This is posing a threat to the Ichthyofauna of North-Western Himalaya. Unless remedial steps are taken to check the trend we may soon lose some of the finest North-Western Himalayan fresh water fish fauna.

References

Buolanger, G.A., 1911. *Catalogue of the freshwater fishes of Africa in British Museum (Nat. Hist.)*, London, 2: 1–529.

Das, S.M., 1960. High altitude fisheries with special reference to Kashmir. *Curr. Sci.*, 29: 220.

Das, S.M. and Subla, B.A., 1963. Ichthyofauna of Kashmir. Part I: History, origin, ecology and distribution. *Ichthyologica*, 2(1–2): 87–116.

Das, S.M. and Subla, B.A., 1966. Plaaearctic elements in the fauna of Kashmir. *Nature*, London, 212: Dec. 17.

Das, S.M. and Subla, B.A., 1970. The Pamir Kashmir theory of origin of Ichthyofauna of Kashmir. *Ichthyologica*, 10 (1–2): 8–15.

Hora, S.L., 1953. Fish distribution and Central Asian prography. *Curr. Sci.*, 22:93–97.

Hora, S.L., 1937. Comparison of the fish fauna of the Northern and the Southern faces of the great Himalayan range. *Rec. Ind. Mus.*, 39: 241–250.

Hora, S.L. and Mukherji, D.D., 1935. Fishes collected by the third Netherland Korakoram expedition. *Wiss Ergab. Neidard., Korakoram*, 1: 426–445.

Hora, S.L., 1955. The evolution of the Indian torrential environment and its fishes. *Bull. Nat. Inst. Sci., India*, 7: 264–268.

Hora, S.L., 1955. Place of Kashmir in the fish geography of India. *Everyday Sci.*, 3: 36–45.

Jayaram, K.C., 1974. Ecology and distribution of fresh water fishes, amphibia and reptiles. In: *Ecology and Biogeography in India*. Dr. W. Junk B.V. publishers, The Hague.

Jayaram, K.C., 1966. Contribution to the study of the Bagrid fishes. *Int. Rev. Ges. Hydrobiology*, 51: 433–450.

Jayram, K.C. and Majundar, N., 1964. On a collection of fish from Kamang frontier division NEFA. *J. Bomb Nat. Hist. Soc.*, 61: 264–280.

Kaul, B.L., 1972. *Unpublished Ph.D. Thesis*, University of Kashmir, Srinagar (India).

Kaul, B.L., 1995. Threat to biogeogephically significant ichthyofauna of Kashmir Himalayas. In: *Ecodegradation of Himalayas*, (Ed) B.L. Kaul. Vinod Publishers, Jammu.

Mani, M.S., 1974. *Ecology and Biogeography in India*. Dr. W. Junk D.V. Publishers, The Hague.

Chapter 11

Morphological and Physico-chemical Properties of Acid Soils from Jammu Himalayan Region (J&K State) and their Management

R.D. Gupta, Daisy Bhan and Deepak Kher

ABSTRACT

Morphological and physico-chemical properties of acidic soils of Jammu Himalayan region, Jammu and Kashmir were studied. It was found that out of 21 soil profiles studied, 2,3, 7, 9, 10, 11, 13 and 19 were very shallow (< 50 cm) and 1,5,6,8,15,16 and 17 shallow in depth (60 to 100 cm), and the remaining were moderately deep (> 100 cm). A lot of variation was seen not only in soils of various profiles but also within the soils of same profile. Generally, sub-surface horizons had finer texture than those of surface. The structure of most of the surface soils was medium/very weak, fine granular and that of sub-surface soils medium moderate granular to the medium moderate sub-angular blocky. Irrespective of soil depth, values of pH_w ranged from 4.5 to 7.0 and those of pH_{KCl} from 4.1 to 6.8 with generally lower values of pH_{KCl} than those of pH_w. In surface horizons, the values of organic carbon ranged from 0.6 to 2.8 per cent and in those of sub-surface from 0.3 to 2.6 with averages of 1.8 and 1.3 per cent. Cation exchange capacity of the studied soils was mainly found to be controlled by the amount of organic carbon. Exchangeable Ca^{2+} was the dominant cation followed by Mg^{2+} and K^{+}. There was great variability in the genesis and classification of the studied soil profiles depending upon topography, parent material, climate and vegetation.

Various methods to manage or reclaim the acidic soils and suggestions/priorities for the future, have also been detailed in this chapter.

Introduction

Acidic soils, which are characterised by low pH *i.e.*, less than 7.0 theoretically and/or less than 6.5 practically. Such soils are deficient in available or exchangeable nutrients especially Ca, Mg and to some extent K. These soils fairly contain high concentration of Fe and Al in the soil solution causing, thereby, fixation of phosphate ions (H_2PO_4, HPO_4). This phosphate fixation phenomenon results in deficiency of available P for the growing plants. Moreover, there becomes toxicity of Fe and Al in such soils. This is more so in case of soils having pH around 5.0 and less than 5.0. Micronutrients like B and Mo also render deficient in acid soils through their fixation mechanism like P by the excessive amount of Fe and Al. Owing to the presence of poor status of available plant nutrients (Ca, Mg, P, B and Mo) in acidic soils accompanied with poor microbial activities (low population of bacteria including nitrifiers and nitrogen fixers, and actinomycetes and higher that of fungi) and toxicity of Fe and Al, the crop production is badly affected.

In view of the facts as stated above it becomes an imperative to study the acidic soils in details not only in respect of available plant nutrients but their morphological, physical and chemical properties also. This will assist in delineating the extent of these soils and, thereafter, to recommend the management practices for their reclamation. There exists a lot of information with regard to the genesis, various characteristics and management of acid soils from various States of India (Biswas, 1976; Mahapatra *et al.*, 1997). Recently, distribution and extent of acid soils in different parts of India, their production constraints, management and crop response to liming and fertilization have appeared in literature (Sharma and Sarkar, 2005) as well as management of acidic soils and their current issues and policy guidelines (Sarkar and Sharma, 2006). There is, however, scanty information available regarding the occurrence of Himalayan acid soils of Jammu and Kashmir, nature of acidity vis-à-vis their characterisation in respect of morphological and physico-chemical properties (Kaul, 1956; Gupta, 1967; Gupta *et al.*, 1992). The present study, therefore, relates to this angle in case of soils of Jammu Himalayan region and their management.

Materials and Methods

A Brief Background of the Study Area

Physiography

The Jammu region of Jammu and Kashmir State partly plain and partly hilly, arises from the Punjab plains with elevation below 300 m and a gentle slope of 3° touching through a rugged topography and bare steep slopes to Outer Himalayan Range. Outer Himalayan Range covers a number of Duns to its north-east, locally called, *Dadhals* (Dansal, Jajhar Kotli, Udhampur, etc.) and Pir Panjal Himalayas to its north side having 3000 to 5000 m altitude with steep escarpments and dip slopes, covering a number of spurs, dales and valleys. Among them, Bhadarwah, Paddar, Loran, Poonch and Surankote are the most leading ones.

Thus, the study area represents outer hills and middle to high mountains of Jammu region, comprising parts of Kathua, Udhampur, Doda, Rajouri and Poonch districts with elevation ranging from 1500 to 2000 m where agriculture is practised (Bali *et al.*, 1997; Bhan, 1999).

Geology

Geologically, the region is composed of large variety of rocks belonging to Siwalik system including Murree series, Jurrassic series, Panjal volcanic series, Zewan series and those representing of Triassic period. Besides, granites of various kinds are also found in parts of Doda district (Gupta and Banerjee, 1991). The rocks of Siwalik system consist mainly of sandstones, silt and claystones, conglomerates and have been further divided into lower, middle and upper Siwaliks. The rocks of Panjal volcanic and Zewan series comprise mainly of Panjal traps, agglomeratic slates, limestones, dolomitic limestones and shales.

Climate

The climate of the region varies from sub-tropical to sub-temperate/temperate and cold arid type (parts of Doda district adjoining to Ladakh region). In the sub-tropical zone, the summer usually starts from April and lasts upto June. June is the hottest month with usual maximum temperature ranging from 40 to 43° Celsius and minimum temperature ranging from 19.5 to 4.5° Celsius. In temperate and sub-temperate regions, the summers are cool and winters are severe cold with usual snowfalls. January and February are the typical cold months with temperatures falling below freezing point. During April to June, the temperature never goes beyond 25° Celsius. The rainfall of whole of the region lies between 1000 mm and 2000 mm, the bulk of which is received from July to September (during monsoon).

Vegetation

The Jammu region is endowed with a variety of flora ranging from subtropical forests to temperate evergreen coniferous and deciduous forests. The main species of subtropical forests are various species of *Acacia* (*A. nilotica, A. modesta, A. catechu*), *Albizzia lebbeck, Albizzia chinensis, Cassia* species (*C. fistula, C. tora*), *Dalbergia sisso, Butea monosperma* and many others. *Pinus roxburghii, Pinus excelsa, Cedrus deodara, Quercus dilatata, Quercus semecarpifolia* are the main species which are found in the subtemperate/temperate moist ares. The main shrubs of the sub-tropical area are *Dodonea viscosa, Vitex negundo, Carissa spinarum, Adharoda vasica, Murraya koenigii* while *Caesalpinia sepiaria* (*Kaudari*), *Berberis lycium* (*Rousant*), are found mainly in sub-temperate and temperate area.

Cropping Pattern

About 26 per cent of the sown area of Jammu region is irrigated. In the outer hills, the cultivation is mainly dependent on rains. Rice, maize and wheat are the main crops which occupy nearly 75 per cent of the net area sown. Pearl millet is grown in *Kandi* belt of Jammu. Other crops grown are oil seeds, pulses and barley.

Collection of Soil Samples and Methods of Analysis

Based on Reconnaissance Soil Survey, 402 surface soil samples were collected from the locations experiencing high rainfall, low temperature and having coniferous type of vegetation. These were analysed for pH values. Out of these soils, only 121 were observed to be acidic having pH in the range of 4.5 to 6.5. Depending upon the similarity of pH, 21 soil profiles were selected and accordingly samples were collected for further analysis. The morphological characters of these soils were studied following Soil Survey Division

Staff Manual (1995). Physico-chemical properties of processed soils were determined using standard procedures (Piper. 1966; Black, 1973; Jackson, 1979). The soils were then classified according to Soil Taxonomy.

Results and Discussion

Morphological Characteristics

The morphological characteristics of different soil profiles in respect of colour, texture and structure vis-à-vis important site characteristics are presented in Table 11.1. Out of 21 soil profiles studied, 2, 3, 7, 9, 10, 11, 13 and 19 were very shallow (< 50 cm) and 1, 5, 6, 8, 15, 16, 17 shallow in depth (60 to 100 cm), and the remaining were moderately deep (> 100 cm).

A lot of variation in colour was noticed not only in soils of various profiles but also within the soils of same profile (Table 11.1). Soil profiles developed at higher elevations *viz.*, 1–3, 5, 8, 9, 11–13 and 15 were generally characterised either by dark brown to strong brown (7.5 YR 4/4 to 7.3 YR 5/6) or black to dark grey (10 YR 2/1 to 10 YR 4/1) and grey to brown (10 YR 6/1 to 7.5 YR 4/3) colour. Dark grey, very dark grey and/or black colour of soils (profiles 2, 3, 5, 9, 11, 12 and 15) could be attributed to coagulation of iron and/or calcium ions with humic substances. These results are in consonance with earlier reports (Gupta, 1992; Najar, 1993). Brown to dark brown colour in horizons of some soil profiles (1, 13, 19, 20 and 21) indicated better drainage conditions whereas pale yellowish brown or dark yellowish brown colour in soil profiles (4, 6, 7, 10, 14, 16–18) revealed poor drainage.

Generally, subsurface horizons had finer texture than those of surface. Soils of profiles 1 to 5, 8, 9 and 11 had loam or silty loam texture, while in rest of the soil profiles it was either sandy clay loam to loam or clay. Great variation in texture could be explained due to difference in parent material from which these soil profiles owed their origin to. However, role of differential colluvium brought about by gravity in hilly terrain and deposited in gently sloping areas could not be ruled out in bringing about variation in texture.

The structure of various soils varied from granular to angular blocky and subangular blocky with rapid to very rapid and slow to very slow permeability. Generally, the structure of most of the surface soils was medium weak/very weak, fine granular and that of

Table 11.1: Morphological and Site Characteristics of Acidic Soils of Jammu Region

Profile No.	Location	Physiography		Vegetation	Rainfall (mm)	Horizons	Range of Morphological Characters		
		Elevation (m)	Slope class Symbol				Colour	Texture	Structure
1.	Batote	1700	E	Deodar forest	1700	4	7.5 YR 4/4 to 7.5 YR 4/3	SL to L	mwfg/m²wg to msbk
2.	Shahdara (upper reaches)	1750	D	Chil, Kail forest	1500	3	10 YR 2/1 to 10R 4/1	L to CL	mwfg
3.	Sansar	1650	E	Potato farm	1650	3	10 YR to 10 YR 4/3	L	mwfg/m²wf to msbk
4.	Sander	520	C	Cultivated	1375	5	7.5 YR to 10 YR 4/6	L	m²g to m² sbk
5.	Bani	1500	E	Forest (mixed)	1625	3	7.5 YR to 7.5 YR 4/3	L	mwg to cwg
6.	Ghat	1050	C	Forest (mixed)	1375	4	10 YR 3/4 to 10 YR 4/6	SiL to CL	wg to cwg/wsbk
7.	Bhoond	1000	C	Forest (mixed)	1200	3	10 YR 3/4 to 10 YR 5/6	L to SL	mwg to wabk
8.	Dahl	1500	B	Cultivated	1500	3	10 YR 4/4 to 10 YR 5/8	L	mwfg to m²sbk
9.	Sanasar	1800	E	Cultivated (potato farm)	1650	3	10 YR 2/1 to 10 YR 3/2	L	wvfg to m²sbk

Contd...

Table 11.1–Cond...

Profile No.	*Location*	*Physiography*		*Vegetation*	*Rainfall (mm)*	*Horizons*	*Range of Morphological Characters*		
		Elevation (m)	*Slope class Symbol*				*Colour*	*Texture*	*Structure*
10.	Mackhadi	1500	B	Cultivated	1375	3	10 YR 4/3 to 10 YR 4/6	CL	wmg to cm^{2}abk
11.	Loran	1950	E	Forest (mixed)	1450	3	10 YR 2/1 to 10 YR 4/1	LSL	wmg to cm^{2}sbk
12.	Loran	1950	0	Forest (mixed)	1450	3	10 YR 2/1 to 10 YR 5/4	L to CL	wmg to m^{2}sbk
13.	Shahdara	1700	C	Oak forest	1500	3	1 0 YR 5/1 to 10 YR 5/4	L to CL	m^{2}sbk
14.	Parnala	510	C	Chil forest	1375	5	10 YR 4/4 to 10 YR 4/6	CL to C	m^{2}abk
15.	Bani	1500	B	Cultivated	1625	3	7.5 YR 4/4 to 10 YR 4/1	SL to CL	mwg to mwabk
16.	Krimshi	800	C	Girna forest	1450	3	10 YR 4/4 to 7.5 YR 5/3	L to CL	mwabk
17.	Pungara	800	C	Shisham forest	1400	4	10 YR 3/3 to 10 YR 5/4	L to CL	mfg to mfgabk
18.	Pungara	800	B	Cultivated	1400	4	10 YR 4/4 to 10 YR 4/6	C to CL	cm^{2}sbk to abk/bk

Contd...

Table 11.1–Cond...

Profile No.	Location	Physiography		Vegetation	Rainfall (mm)	Horizons	Range of Morphological Characters		
		Elevation (m)	Slope class Symbol				Colour	Texture	Structure
19.	Kanthgali	900	C	Chil forest	1450	5	7.4 YR 2/0 to 7.5 YR 5/4	CL	vfg/fg to sbk/bk
20.	Pathi Drekara	750	B	Cultivated	1350	3	10 YR 5/3 to 2.5 YR	L to CL	cm²sbk to bk
21.	Pathi Drekara	750	B	Chil forest	1350	5	5/3 to 2.5 YR 5/4	L to CL	fg to sbk/bk

Note: Symbols of morphological characteristics followed are those of Soil Survey Manual [Soil Survey Staff, 1995] and slope Classes of Sehgal *et al.* (1987). B: Very gently sloping-1 to 3 per cent; C: Gently sloping; D: Moderately sloping: 8 to 15 per cent; E: Moderately steep sloping–15 to 30 per cent.

subsurface soils medium moderate granular to the medium moderate subangular blocky. The moderate medium weak granular to coarse weak granular/weak granular or medium weak fine granular structure in soils of profiles, 4–6, 8, 10–12 and 15, pointed out their recent origin. Whereas, cm^2 sub-angular blocky/blocky or angular blocky/blocky structure in subsurface and surface horizons of 18 and 20 exhibited structural development. Fine granular structure in most of surface soils might be due to presence of high buildup of organic matter in them. Wide variation in permeability could be elucidated because of differences in sand and/or clay contents.

Physico-chemical Properties

The data with regard to physico-chemical properties *viz.*, pH_w, pH_{KCl}, organic carbon, CEC and exchangeable bases (Ca^{2+}, Mg^{2+}, K^+) and mechanical components (sand, silt, clay) of various soil profiles have been presented in Table 11.2. Irrespective of soil depth, values of pH_w ranged from 4.5 to 7.0 and those of pH_{KCl} from 4.1 to 6.8. In surface horizons pH_w ranged from 4.5 to 6.8 and subsurface horizon from 4.9 to 7.0, with average values of 5.7 and 5.8, indicating slightly higher pH_w in subsurface horizons. It is point to mention that pH_{KCl} was generally lower than pH_w, the range values being 4.5 to 6.4 and 4.4 to 6.8 with average values of 5.2 and 5.3. Soil profile 1 from Batote under deodar forest had the lowest pH values ranging from pH_w–4.5 to 5.3 and pH_{KCl} to 4.1 to 4.7 (Table 11.1). Increase in pH_w with depth in soil profiles 1 to 10 and 13 (Bhan, 1999) could be ascribed possibly to leaching of bases from the surface to subsurface. However, higher pH_w values in surface soils of other profiles than those of subsurface suggested an accumulation of bases on surface through vegetation. Low pH_w in soil profile of Batote was possibly due to high rainfall (Table 11.2). Influence of rainfall on soil pH has also been reported by Singh *et al.* (1993). However, role of humified nature of organic matter in terms of humic and fulvic acids in lowering the pH in this profile could not be ruled out as they produce more exchangeable Al^{3+} and Fe^{3+} from the primary/clay minerals by weathering which in turn lower down the pH of the soils. On the basis of pH rating limits fixed by Mandal *et al.* (1975), the studied soils have been classified into strongly acidic (profile 1), moderately acid (2, 5, 6) and acidic (3, 4, 7, 8, 11, 12, 13, 14, 15, 16, 17, 19, 20, 21) and slightly acidic (9,10,18). Their details have been shown in Table 11.3.

Table 11.2: Physico-chemical Characteristics of Acidic Soils of Jammu Region

Profile No.	pH_w	pHKcl	OC (%)	CEC and Exchangeable Bases [Cmol (p^+) kg^{-1}]				Base Saturation (%)	Mechanical Components (%)		
				CEC	Ca^{2+}	Mg^{2+}	K^+		Sand	Silt	Clay
1.	4.5–5.3	4.1–4.7	1.6–2.8	9. 1–9.9	4.1–4.9	1.5–1.8	0.11–0.20	69–70	30–46	40–58	12–18
2.	5.0–6.0	4.6–5.0	2.4–2.8	9.8–10.5	5.7–6.0	1.5–2.0	0.15–0.20	78–79	27–38	37–41	22–36
3.	5.2–6.2	4.7–5.5	1.4–1.8	7.0–7.2	3.8–4.0	1.0–1.2	0.11–0.14	72–74	40–44	32–37	23–24
4.	5.2–6.5	4.7–5.7	0.0–1.8	6.4–6.8	3.5–3.9	0.6–0.8	0.1 0–0.14	70–75	33–44	38–44	18–27
5.	5.2–5.7	4.7–5.0	1.0–2.3	9.5–10.4	5.1–5.8	1.8–1. 9	0.07–0.09	76–77	29–40	38–46	21–28
6.	5.2–5.6	4.6–4.8	1.2–2.9	10.3–10.3	6.3–6.9	1.6–1.8	0.11–0.15	80–85	29–54	24–34	22–38
7.	5.5–6.0	4.9–5.5	0.6–2.0	9.1–9.8	5.0–5.7	1.8–1. 9	0.06–0.07	79–81	27–38	43–51	18–30
8.	5.5–6.2	4.9–5.8	1. 0–1.9	7.0–7.2	3.6–3.9	1.3–1.5	0.16–0.20	73–80	36–44	34–40	22–24
9.	5.5–5.8	4.9–5.1	1.5–2.3	7.9–8.1	3.8–4.2	1.6–1. 9	0.11–0.14	74–75	44–47	33–37	19–20
10.	5.7–6.8	5.0–5.9	0.5–1.8	8.5–9.2	5.5–5.7	1.0–1.3	0.16–0.34	79–82	30–44	27–34	26–36
11.	5.6–5.8	4.9–5.2	1.9–2.4	8.3–8.9	4.7–4.9	1.6–1.8	0.11–0.12	78–80	27–38	44–53	18–20
12.	5.6–5.8	5.0–5.1	1.0–2.0	8.4–8.8	4.5–4.9	1.7–2.0	0.10–0.11	78–83	25–35	45–51	20–30
13.	5.9–6.2	5.1–5.6	1.5–2.0	8.9–9.4	4.7–4.9	1.9–2.1	0.10–0.12	76–77	30–36	38–44	20–32
14.	5.5–6.1	5.0–5.7	0.5–1.1	9.4–9.9	5.7–6.0	1.9–2.1	0.11–0.14	85–86	34–50	20–22	30–45

Contd...

Table 11.2–Contd...

Profile No.	pH_w	pHKcl	OC (%)	CEC and Exchangeable Bases [Cmol (p^+) kg^{-1}]				Base Saturation (%)	Mechanical Components (%)		
				CEC	Ca^{2+}	Mg^{2+}	K^+		Sand	Silt	Clay
15.	5.3–7.0	4.4–6.8	0.3–0.7	5.4–7.1	3.2–4.2	0.7–0.9	0.10–0.16	71–85	28–67	20–43	13–29
16.	5.8–6.7	5.1–6.1	0.8–1.4	5.7–13.2	3.1–7.4	1.2–3.6	0.13–0.38	65–90	26–34	44–48	18–30
17.	5.8–6.2	5.1–5.8	0.9–1.6	5.7–8.5	2.7–5.3	0.6–1.0	0.07–0.12	66–74	28–44	21–34	21–35
18.	6.3–6.6	5.8–6.0	0.5–0.6	4.5–6.6	2.0–4.0	0.7–1.1	0.06–0.11	68–74	37–46	22–25	33–50
19.	5.8–6.2	5.1–5.8	1.0–2.6	5.9–11.0	1.5–6.5	1.3–2.3	0.15–0.22	62–78	30–40	29–39	21–40
20.	6.4–6.8	6.0–6.4	0.8–0.9	6.1–6.3	2.5–2.7	1.6–2.1	0.16–0.23	74–83	34–44	34–37	22–29
21.	5.5–6.6	5.0–6.1	0.8–2.2	6.3–9.9	2.0–3.8	1.0–2.5	0.07–0.23	67–78	25–46	33–38	20–37

Table 11.3: Details of Acidic Soils of Present Study

Sl.No.	Soil Group	pH Range	Area of Studied Profiles
1.	Strongly acidic	4.0 to 5.0	Batote under forests having undulating topography.
2.	Moderately acidic	5.1 to 5.5	Shahdara (upper reaches), Bani and Ghat under forests.
3.	Acidic	5.6 to 6.0	Sanasar, Sander, Bhoond, Loran, Shadara (lower reaches), Parnala, Pungara, Bani cultivated, Kanthgali, Mansar, Krimchi and Pathi Drekara.
4.	Slightly acidic	6.1 to 6.5	Macchadi, Pungara and Bani (cultivated).

Source: Bhan, 1999.

Organic carbon (OC) content in various soil profiles ranged from 0.33 to 2.8 per cent (Table 11.2) with an average of 1.5 per cent. In surface horizons, its value ranged from 0.6 to 2.8 per cent and in those of subsurface from 0.3 to 2.6 with averages of 1.8 and 1.3 per cent, indicating more values in surface horizons. Higher values of OC in surface soils of all profiles (Bhan, 1999) were due to the addition of organic matter through forest litter, organic residues and cover crops and their subsequent humification. More organic carbon content in soil profiles of Batote, Ghat, Loran, Shahdara at higher elevations were ascribed to slow rate of organic matter decomposition at low temperature prevailing in these areas. These findings were more or less in agreement with earlier reports (Gupta and Tripathi, 1992 and Gangopadhyay *et al.*, 1998).

CEC of the soil profiles varied from 4.5 to 13.2 Cmol (p^+) kg^{-1} with an average of 8.2. Generally, there were slightly [5.1 to 8.5 Cmol (p^+) kg^{-1}] higher values of CEC in surface soils than those of subsurface [4.4 to 10.8 Cmol (p^+) kg^{-1}]. Although CEC values in soils of present study were low as compared to earlier reports (Gupta and Verma, 1992) yet they varied with OC and clay content (Bhan, 1999). However, the CEC values were not very conspicuous with clay content. This indicates that CEC of these soils was controlled mainly by the amount of OC which was further supported by highly significant positive correlation of this parameter with OC ($r = +0.744^{**}$). Higher values of CEC in surface than subsurface soils were explainable on the basis of more OC in the former

(1.80 per cent) than the latter (1.36 per cent). Low values of CEC in the soils of present study than those obtained by the aforementioned workers were due to the difference in the methodology used. In the present study, CEC was worked out by summation of exchangeable bases rather than determined by distillation of ammonical N after leaching the soils with normal ammonium acetate, the values of which are generally more.

Exchangeable Ca^{2+} was predominant ranging from 1.5 to 7.4 Cmol (p^+) kg^{-1} followed by Mg^{2+} and K^+, the amount of which varied from 0.6 to 2.5 and 0.06 to 0.34 Cmol (p^+) kg^{-1}, respectively. These findings were in conformity with earlier reports (Raychaudhuri *et al.*, 1963; Gupta, 1992), according to which the soils of Jammu and Kashmir were rich in exchangeable Ca^+ and Mg^{2+}. Mg^{2+} as second important cation after Ca^{2+} in these soils was due to the presence of chlorite type of clay mineral which has Mg^{2+} in its brucite layer (Gupta, 1967). Slightly more amount of exchangeable Ca^{2+} and Mg^{2+} in surface soils was due to more weathering of soil particles, especially in organic colloids than those in subsurface. However, role of organic matter in release of these cations during its decomposition could not be ruled out. The difference in cations in soils of various profiles were due to variation in physiographic position of the profiles, parent materials, rainfall and vegetation as stated by Gangopadhyay *et al.* (1998).

Textural variations (Bhan, 1999) were attributed to *in situ* weathering of soil profiles under different types of vegetation, parent material and rainfall conditions (Verma *et al.*, 1987; Walia and Chamuah, 1992; Gupta and Tripathi, 1996). On an average more clay content in subsurface soils (28.5 per cent) than those of surface (21.7 per cent) suggested its downward movement. Similar observations have been made by Maji *et al.* (1998).

Genesis and Classification

The morphological and physico-chemical properties of soils of different profiles (Tables 11.1 and 11.2) showed striking variability in the genesis of these soils. Except soils of profiles 2 and 12, soils of profiles 14, 16, 17, 18, 19 and 21 have developed on very gently sloping to gently sloping topography. Thus, there is regular and considerable increase in clay content with depth fulfilling the requirement of an argillic horizons. Soils of profiles (2 and 12) though have developed under moderately sloping topography yet they support forest vegetation which is corroborated by more amount of organic carbon, resulting into alluviation of clay. As there were no mottlings in the ped faces of the soils of these 8 profiles and as such were classified into Typic Paleudalf subgroup level (Alfisol order; Udalfs suborder and Paleudalf great group levels).

Soils of profiles 6, 7 and 13 are forest soils and have developed on gently sloping (3–8 per cent) topography. They are stable and undisturbed landscape. Hence, they are well developed owing to the presence of an argrillic horizon and placed in Alfisol order, Udalfs suborder, Hapludalfs great group and Typic Hapludalfs subgroup level.

Profiles 1, 10 and 15 have been classified into Typic Udifluvents subgroup level under Entisols order, Fluvents suborder and Udifluvents great group levels. In case of profile 1, there is, of and on, occurrence of landslide in this profile though it is under forest vegetation. This phenomenon accompanied with moderately steep sloping (15–30 per cent) topography have hindered the development of this profile. In case of profiles 10 and 15 having very gently sloping (1 to 3 per cent) topography, there appears to be some old water transported or fluvial deposits in these cultivated profiles. However, youngsters of the area did not know much about this event and the elders were not available at the time of collection of soil samples. Hence, this event could not be confirmed.

Profiles 3 and 9 developed on moderately steep sloping (15–30 per cent) topography have got less chance to profile development and hence have been put into Entisols order, Orthents suborder, Udorthents great group and Typic Udorthents subgroup level. Moreover, being situated on potato farm, it seemed that terracing might have undertaken in the recent past which disturbed the genetic horizonation of these soil profiles. Profiles 5 and 11 are also developed on steep sloping (15–30 per cent) topography and have obtained less chance to develop due to steepness. Profile 8 though developed on very gently to gently sloping (1–3 per cent) topography yet the alluvium parent material brought by the stream near the profile, has created obstacles in the genesis of this profile. Therefore, the profiles 5, 8 and 11 have been classified into Typic Udorthents subgroups levels like those of 3 and 9.

Profile 4 on gently sloping (3 to 8 per cent) topography has shown beginning in the soil profile development, so it belongs to

Typic Eutrochrepts subgroup under the Inceptisols order, Ochrepts suborder and Eutrochrepts great group. Profile 20 under cultivation and situated on very gently to gently sloping (1–3 per cent) topography, has also shown beginning in its development. Since the profile is located near the stream, so it has been put under Fluventic Eutrochrepts subgroup level under the Inceptisols order, Ochrepts suborder, Eutrochrepts great group level.

Management of Acidic Soils

Response of Crops to Liming

To reclaim or manage the acidic soils, application of liming has been found to be the best method to control acidity. Its application not only helps in checking soil acidity but also assists in increasing crop production. The crops which have shown much more response belong to legumes, followed by those of cotton, maize, sorghum, wheat and linseed (Sarkar and Sharma, 2006). In addition to these crops, promising results of liming have also been obtained in case of *Phaseolus mungo* in Bihar, jute and mustard in Assam, jute in West Bengal, horse-gram in Orissa, paddy in Karnataka and wheat and maize in Himachal Pradesh.

In Jammu and Kashmir, aforesaid studies are lacking. It is because not much work has been carried out so far regarding the occurrence and/or presence of acidic soils in compact area. It. therefore, calls for attention to agricultural scientists working both in the State Agricultural Universities and, Agriculture and Horticulture Departments to start with this work in view of the occurrence of acidic soils as observed in the present study.

Role of Liming

Liming is a widely accepted practice for improving an efficient utilization of native soil nutrients. As a matter of fact, application of lime increases the soil pH by way of reducing exchange acidity and increasing the base saturation. Increased base saturation inactivates Fe and Al concentration and, thus, reduces P fixation largely, stimulates the microbial activity leading to mineralisation of organic matter.

Sources of Liming

Powder of limestone and dolomitic limestone has been used successfully since long as an amendment to manage the acidic soils to improve crop productivity. However, economic consideration is the most vital factor to popularise any practice among the farmers. Liming material must have low cost for use by small and marginal farmers. It should be locally available.

There are large resources of limestone in number of States like Arunachal Pradesh. Manipur, Assam, Meghalaya, Nagaland and Jammu and Kashmir. Paper mill sludges, basic slag from steel industry are the other liming materials which can be used to ameliorate the acidic soils. Jharkhand State has lot of resources in respect of basic slag and dolomite.

Technology for Acidic Soil Management

Lime application at the rate of 2-4 qha^{-1} has been recommended in many States to reclaim rainfed and irrigated uplands having soil pH < 5.5 (Sarkar and Sharma, 2006). It should be applied in powder form in furrows alongwith the basal dose of fertilizers (part of nitrogenous fertilizer and full dose of phosphatic and potassic fertilizers) manually or through seed-cum-fertilizer drill at the time of sowing.

Suggestions and Priorities for the Future

1. There is no doubt that the present study vis-à-vis that which has been detailed elsewhere (Bhan and Gupta, 2001), have brought few new facts into light. For example, the presence of strongly and moderately acidic soils (pH ranges from 4.0 to 5.0 and 5.1 to 5.5). Of the various forms of acidity studied, exchangeable pH dependent and potential acidity, the per cent contribution of pH dependent acidity towards the potential acidity was 83.3.
2. Aforesaid studies, however, have not been conducted so far for the soils of Kashmir Himalayas. It, therefore, becomes imperative to undertake such studies on priority in these soils.
3. Soil mapping upto block level in both the provinces or divisions of Jammu and Kashmir State to identify the acidic problematic areas is another study which needs to be conducted.
4. In mapping the, problematic acidic soils, the staff of Chief Soil Survey Officers of both the Divisions as well as that of

Soil Testing Laboratories must come forward in this aspect in collaboration with the scientists of SKUAST-J and SKUAST-K.

References

Bali, S.V., Gupta, D.K. and Gupta, R.D., 1997. Integrated nutrient management for the farmers of Jammu province. *A SKUAST Bulletin*, Jammu.

Bhan, D., 1999. Characterisation of acid soils of Jammu region. *M.Sc. Thesis*, Division of Soil Science, FOA, SKUAST, J&K, Srinagar, India.

Bhan, D. and Gupta, R.D., 200.1 Forms of acidity in some acid soils of Jammu (J&K). *SKUAST J. Res.*, 3: 21–25.

Biswas, T.D., 1976. Acid soils of India, their genesis, characteristics and management. *Bulletin J. Indian Soc. Soil Sci.*, 11: 1–308.

Black, C.A., 1973. *Soil Plant Relationship*. Wiley Eastern Limited, New Delhi.

Gangopadhyay, S.K., Walia, C.S., Chamuah, G.S. and Baruah, 1998. Rice growing soils of Upper Brahmaputra Valley of Assam: Their characteristics and suitability. *J. Indian Soc. Soil Sci.*, 46: 103–109.

Gupta, R.D., 1967. Genesis, physical, chemical, mineralogical and microbiological nature of the soils of Jammu and Kashmir State. *M.Sc. Thesis*, Department of Agricultural Chemistry, Ranchi Uni., RAC, Kanke, Ranchi, India.

Gupta, R.D., 1992. Soil genesis in the Jammu and Kashmir Himalayas. In: *Himalayan Environment: Man and the Economic Activities*, (Ed.) J.L. Raina. Pointer Publishers, Jaipur, Rajasthan, India.

Gupta, R.D. and Banerjee, S.K., 1991. *Problems and Management of Soils and Forest Resources of North West Himalayas*. Mahajan Book Centre, Last Morh, Gandhi Nagar, Jammu, J&K.

Gupta, R.D. and Tripathi, B.R., 1992. Genesis of soils in wet temperate and subalpine/moist alpine climatic zones of North West Himalayas. *J. Indian Society Soil Science*, 40(3): 505–512.

Gupta, R.D. and Tripathi, B.R., 1996. Mineralogy, genesis and classification of soils of North West Himalayas develped on different parent material and variable topography. *J. Indian Society Soil Science*, 44(4): 705–712.

Gupta, R.D. and Verma, S.D., 1992. Characterisation and classification of some soils of *Kandi* belt from Jammu Siwalik Hills. *J. Indian Society Soil Science*, 40(4): 809–815.

Gupta, R.D., Gupta, J.P. and Singh, H., 1992. Problems of agriculture in the Kashmir Himalayas. In: *Conserving Indian Environment*, (Ed.) S.K. Chadha. Pointer's Publishers, Jaipur, India.

Jackson, M.L., 1979. *Soil Chemical Analysis*. Prentice Hall of India Pvt. Ltd., New Delhi.

Kaul, M.L., 1956. Study of soils at different altitudes in Kashmir State. *Thesis*, IARI, New Delhi.

Mahapatra, I.C., Mandal, S.C., Misra, C., Mitra, G.N. and Panda, N., 1997. *Acid Soils of India*. ICAR, Krishi Anusandhan Bhavan, New Delhi, India.

Maji, B., Bandyopadhyay, B.R., Sarkar, D. and Chaterji, S., 1998. Morphological and chemical characteristics of soils of Sagar island of the Sundarbans, West Bengal. *J. Indian Society Soil Science*, 46(1): 99–103.

Mandal, S.C., Sinha, M.K. and Sinha, H., 1975. Acid soils of India and liming. *ICAR Technical Bulletin* No. 51.

Najar, G.R., 1993. Characterisation and management of salt affected soils of Jammu district, J&K. *M.Sc. Thesis*, Division of Soil Science, SKUAST, Srinagar, J&K.

Piper, C.S., 1966. *Soil and Plant Analysis*. Interscience Publications Inc., New York.

Raychaudhuri, S.P., Agarwal, R.R., Datta Biswas, N.R. and Thomas, P.K., 1963. *Soils of India*. ICAR, New Delhi.

Sarkar, A.K. and Sharma, S.P., 2006. Management of acidic soils: Current issues and policy guidelines. In: *Balanced Fertilization for Sustainable Crop Production, Proceedings National Seminar on INM for Balanced Fertilization*, November 17–18, (Eds.) Sharma, S.P., Sharma, S.K., Sharma, R.P. and T.S. Verma. Department of Soil Science, CSK HPKV, Palampur, HP, India.

Sharma, P.D. and Sarkar, A.K., 2005. *Managing Acid Soils for Enhancing Productivity*. ICAR, Krishi Anusandhan Bhavan, New Delhi, India.

Singh, R.N., Diwakar, D.P.S. and Singh, A.K., 1993. A comparative study of acid soils developed on granite gneiss and Mahananda alluvium. *J. Indian Society Soil Science*, 41(1): 125–132.

Soil Survey Division Staff, 1995. *Soil Survey Manual* (Indian Print). Handbook No. 18 (USDA, Washington DC).

Soil Survey Staff, 1999. *Soil Taxonomy. A Basic System of Soil Classification for Making and Interpretation of Soil Survey*, 2nd Edition. USDA (National Resource Conservation Service, Washington DC).

Verma, T.S., Tripathi, B.R. and Verma, S.D., 1987. A comparative study on the nature and properties of soils in relation to climate and parent material. *J. Indian Society Soil Science*, 54(1): 85–91.

Walia, C.S. and Chamuah, G.S., 1992. Flood affected soils of Brahmaputra Valley and their suitability for land use planning. *J. Indian Society Soil Science*, 40: 335.

Chapter 12

Management and Conservation of the Natural Resources of the Himalayas with Emphasis on Kashmir Himalayas

Anil Kumar Verma and B.L. Kaul

ABSTRACT

The Himalayas are the loftiest and greatest of all the mountain systems in the world. They have from time immemorial held an indescribable charm, inspiration and spiritual significance for the millions living in and around these mountains. It is well impossible to describe in words the mysticism and the religious aura that surround these mountains. Great Sages and poets have tried to describe the sheer majesty and charm of the Himalayas but could never fully capture in words their majestic grandeur.

The Himalayan land and water are priceless national assets. The approach of the modern man to explore and exploit its resources is very harsh. If the Himalayan region is over used to meet the increasing demand for fuel, fodder, timber, etc., it would be disturbed and it would create environmental hazards in the Himalayas. The question of conservation, regeneration and development cannot be left to chance and the vagaries of political and administrative setup at the lowest level. Conservation and development need to be made mandatory without prejudice to the rights which are inherent in the land. In this way, objections or dilatory tactics can be given short shrift.

Introduction

Geography of the Himalaya

The Himalayas are an awesome chain of mountains that extend approximately to 2500 km. From north-west to south-east and the width is between 250 to 300 km. The Kashmir Himalayan system consists of three longitudinal belts separated by faults, their geological and topographical characters are strikingly different. Immediately north of the plains is the Siwalik Zone. The width of this zone varies from 10 to 50 km. and altitude rarely exceeds 900m. This zone is covered with damp and scrub forest. The next zone is known as the middle Himalaya with width ranging from 65 to 75 km and the average height being 3000 m.

This zone is profusely dissected by the major rivers passing across it. The third belt is the Great Himalaya rising to an average height of 5,200 m. The great Himalayan peaks are situated in this zone. There are more than 90 peaks having an altitude of more than 8,000 m. This zone has two sub-zones, *viz.*, an alpine zone (4800 m) and an upper snow-bound zone (7500 m).

The transverse sub-divisions of the Himalaya are as under:

1. The Punjab Himalaya: Sutlej to Indus, 590 km.
2. The Kumaon Himalaya: Kali to Sutlej, 350 km.
3. The Nepal Himalaya: Tista to Kali, 800 km.
4. The Assam Himalaya: Brahmaputra to Tista, 750 km.

Dr. Valdiya, an eminent geologist has strongly supported the rise of the Himalaya out of a vast tethys geosyncline that existed where now we see the Himalayas. The soft sediments that lay at the bottom of the geosyncline were pressed, upheaved in several phases. The present configuration and complexity of the structure of the Himalayas is due to the phenomenon of "plate tectonics" and other related orogenic activities. The Himalayas are very young mountains and constitute less than $1/5^{th}$ of the total geographical area of India. They control the climate and through the rivers the prosperity of more than one third of the total population living in the plains of the Indus, the Ganges and the Brahmaputra.

As Reinold Messner wrote recently, "Man, through the centuries, has always chosen places which seemed special and declared them holy and untouchable. There, the gods lived. Today more than ever we have a need of unexplored wildernesses." And our Gods have always lived in the Himalayas-the Crumbling Himalayas.

The Himalayan range stretching from Pakistan to Arunachal Pradesh in India must be one of the most spectacular and dramatic areas on earth. Not only does it embrace the highest mountains in the world, but also mighty glaciers and ferocious rivers, huge foothills, fertile valleys, extensive deserts and dense forests.

And it is not only the topography that is so impressive, there is a considerable variety of mountains people with different cultures, religions and language. Four decades ago tourism, as we know it today, was almost non-existent. And possibly that has been the major change in the Himalayan activity-the development of number and quality of airfields in the mountains. Nowadays tens of thousands of people are walking the hill tracks, floating, landing in tiny airfields in remote valleys, and visiting the sacred places in the mountains.

The impact of all these people visiting the distant mountain areas has been tremendous. Some of the effects are good and some not so good. The standard of living of the mountain people has increased considerably from tourist wealth. But the beautiful tracks and valleys have been littered with rubbish and the forests have been destroyed to provide fires for meals and comfort for tourists. The Himalayan environment has suffered severely, partly from tourism but more importantly from the substantial growth in the local population. Arable land is increasingly rare, the forests are cleaned and the slopes terraced for the planting of crops. Population growth is a massive problem to overcome–nobody really knows how to deal with it.

As the tourist numbers grow, the impact on the most popular areas becomes more and more severe. But in the huge stretch of the Himalayas there are a multitude of beautiful areas and many of these are completely neglected. Energetic efforts should be made to attract people to the less popular places, to spread tourism thinner, so that their customers leave the hills and valleys in as pristine a condition as they were when they found them. One of the great joys of adventure tourism is getting sway from the everyday whirl of living, appreciating the peace and contentment of areas, free from public pressure, enjoying the bubbling streams, quiet forest paths, and lonely mountain spaces. We all need to escape from the concrete jungle and return refreshed to the city environment and dream of sparkling waters and icy mountain tops.

Water Resources

The Himalayas as the name implies are the "abode of snow". The total ice cover of the Himalayas is around 16,000 sq. km and this amounts to 17 per cent of their area. A very large number of glaciers exist at present above the snow line, most of which are in the Himalayan range. Glaciers provide water to major perennial rivers of the Himalayan system. Glaciers are formed from the accumulated snow which has survived melting; for a glacier to exist, the snowfall should exceed the quantity that melts in a summer.

There are around 15,000 glaciers in the Himalayas. They vary in length and majority of them are one to five kilometers long. The Gangetic glacier in the Central Himalaya is the source of the Ganges. Its ice surface occupies little over 20 cu. km of ice. Satopath is another glacier which forms the secondary source of the Ganges. The Brahmaputra river which ranks among the mighty rivers of the sun-continent of India, rises in the Great Glacier in the Kailash range in the northern most chain of the Himalaya just south of Lake Tasho in the Tibetan region.

The Indus, the Ganges and the Brahmaputra are the three major rivers systems having their source in the Himalayas. They drain the country and provide a large potential for the irrigation and domestic use and also for the generation of hydroelectric power. In implementing the development programmes it is necessary to ensure that the adverse environmental impacts do not arise. The total catchment area of these three rivers is about 8,00,000 sq. km. Out of the total utilizable surface water resources of 666 million cu.m. over 245 million cu.m. is accounted by these three mighty Himalayan rivers.

The Himalayas play a vital role in replenishing the water resources of India. They act as a barrier for moisture on the southern slopes. The melting of the snow and enormous rain keep the drainage system of the Himalayas healthy and further enrich the subsurface flow of water. The Himalayan glaciers are the banks of water. Harnessing of the Himalayan water resources so essential for food security of Indo-Gangetic plain should be taken up in such a manner that ecosystem is not disturbed. The productivity of water resources can be enhanced by trout culture in the free floating water in higher altitude areas and by raising the crops in the standing waters.

Forest Resources

The Himalayan region is the largest mountain system in the world, of paramount importance in the context of biodiversity. The traditional utilization of biologically diverse resources in the Himalayas not only reflects a diverse resource use pattern, but also a respect for preserving biological diversity by the mountain people.

Forests undoubtedly are the most important biodiversity resource in the Himalayan Mountain Region. In mountain agriculture both permanent farming and shifting systems are directly or indirectly nurtured by mountain forests. Mountain forests also provide numerous products and are essential for environmental protection. The local people utilize various bio-species found in forests. They manage the environmental resources in their habitat as per their subsistence needs. The native mountain people regard themselves as a part of the ecosystem.

The forest management practices are based on their perception of the man-forest relationship from which they evolved the concept of the 'Holy Hills' *i.e.*, an area of forested hillside where the gods reside. A respect for the gods, goddess, deities, and religious and social beliefs promotes peace and well being of the local tribals of this region. Likewise any disrespect, on the other hand, incurs the wrath of the gods.

The following types of forests are (Figures 12.1–12.3; Table 12.1) found in India which cover nearly 17 per cent of total area of Indian territory.

Table 12.1: Showing Composition of Indian Forests

Sl.No.	*Kinds of Forests*	*Percentage*
1.	Tropical forests	
(*a*)	Deciduous	80 per cent
(*b*)	Evergreen	12 per cent
(*c*)	Others	1 per cent
2.	Temperate forests	
(*a*)	Coniferous	3 per cent
(*b*)	Broad leaved	4 per cent

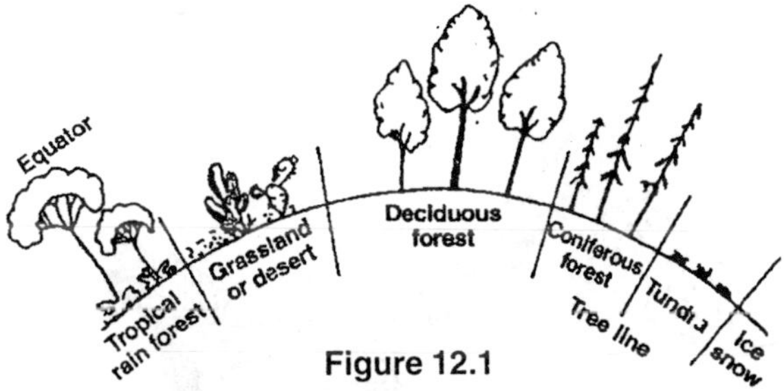

Figure 12.1

Figure 12.2

Figures 12.1 and 12.2: Stratification in a Tropical Rain Forest

The *alpine forests* occur at heights between 2,900 and 3,800 metres in the Himalayan region. Being subjected to heavy snow, the trees here are stunted, mishappen and bent. Several species of Rhododendrons and Junipers are found here. The papery barked (Bhurjapatra) *Betula*, which was used for writing in days gone by, is also found in the alpine forests. In the eastern Himalayas the silver fir (*Abies densa*) is common, *Bambusa arundinaria*, the reed-like bamboo, occurs in places where the undergrowth has been burnt by graziers.

The *sub-alpine forests* stand between 2,700 and 1,800 metres. The deodar (*Cedrus deodara*), a timber tree, and the blue pine, *Pinus wallichiana*, are the important members of these coniferous forests in the dry western region. The blue pine is found at the higher elevations and the deodar at lower levels. The spruce (*Picea smithiana*) is also found along with the blue pine at the upper limits. Lower down there

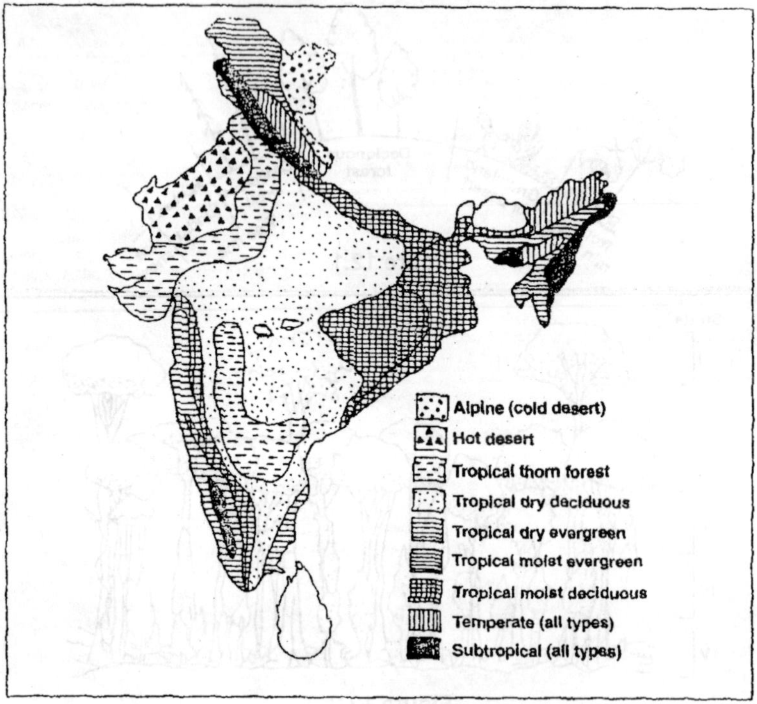

Figure 12.3: Many Kinds of Forests found in India

are several species of broad-leaved trees like the oak (*Quercus*) and others.

In the Eastern Himalayas, Sikkim and Darjeeling, the rainfall is quite heavy. Snowfall is very limited and of short duration; there are more broad-leaved species here like oaks, laurels, maples, hemlock and yew.

The *Himalayan forests* occur at elevation of 750 m. to 1,800 m. These have been under intensive interference by the hill tribes, particularly shifting cultivation. Large areas have been brought under cultivation to grow tea, rubber, coffee, etc. As a result, only a few pockets still retain this type of forest. Subtropical pine forests found in the Khasi, Jainti, Naga and Manipur hills and westward from Darjeeling right upto Kashmir on steep hills which were not cleared. The chir pine (*Pinus roxburgii*) is found in the Western Himalayan region.

The *moist tropical forest* constitutes 300 million hectares (nearly half) of the total forested areas. It is divided into four major subgroups.

The southern *tropical moist evergreen forests* occur in the Western Ghats in Maharashtra, Karnataka and Kerala and in the Andamans. Dipterocarps (*Callophylu, Artocarpus, Dipterocarpus,* etc.) are the predominant trees. A few climbing bamboos and plenty of cane are typical of such forests. The southern tropical and plenty of cane are typical of such forests. The southern tropical semi-evergreen forests grow in close association with the true evergreens; they are found in the Western Ghats, Goa, Kerala and Andaman. *Xylia* and *Terminalias* (Aujuna, Deshi Badam, Hirad, Behada) are the most common varieties interspersed with Dipterocarps in the upper canopy. The lower canopy consists of evergreen varieties like *Myrtaceae* (Guava family), *Lauraceae* (*Laurus, Actinodaphne*) etc. and some bamboos.

The *north tropical moist evergreens* occur throughout north-east India. Shifting cultivation has given rise to bamboos in the cleared areas. Dipterocarps are present all over, except in Bengal. These forests have more varieties than their southern counterpart. The semi-evergreen forests in the deciduous components. Sal (*Shorea robusta*) has come to establish" itself due to fire destruction and dessication.

The *tropical moist deciduous forests* are the most important commercially. Teak (*Tectona grandis*) and sal are the principal trees found in these forests which are distributed in the Andaman, Andhra Pradesh, Orissa, Tripura, Assam, Bihar, Uttar Pradesh, Madhya Pradesh, Gujarat, Maharashtra, Karnataka, Kerala and Tamil Nadu. Several other timber trees are mostly teak forests. *Terminalia,* bamboos and several other trees also grow here.

The *dry tropical forests* are of two types: the *dry dedicuous* and the *dry evergreen.* The deciduous types are found over a wide area in the Deccan plateau around the major rivers. They contain fewer species than the moist dediduous forests. The forests are somewhat open. They are subjected to heavy destruction, clearing for agriculture, felling for fuel, overgrazing and fire. There is dense growth of grass during the rains. *Dendrocalamus striclus,* a type of bamboo, is typical of this region. Sandal (*Santalum album*) is found in Karnataka and Tamil Nadu States.

A small pocket of land in the Peninsula from Tirunelveli (Tamil Nadu) to Nellore (A.P.) has the *dry evergreen forest*. It is subjected to heavy deforestation, and once destroyed, it is difficult to restore. It consists of trees with short boles and spreading crown of *Manileara hexandra, M. elengi, Diospyros* and species of *Eugenia*. The thick undergrowth consists of spiny shrubs and a tangle of climbers.

The poorest type of forests are the *thorn forests* found in the desert regions of Rajasthan and Punjab, Western U.P., Gujarat and parts of the Deccan plateau. Most common here are the Acacias. Thorny species of *Euphorbia, Zizyphus* (Ber), *Mimosa*, etc., among others are present.

Forest Resources in Kashmir Himalayas

The State of Jammu and Kashmir is well endowed in forest resource. Its natural vegetation has great diversity in flora, ranging from the lush-green *Margs* (Alpine pastures) to evergreen conifers on the gentle slopes of high altitudes, and from scrub-jungles of the foot-hills to the deciduous forests on the southern slopes of the Siwaliks and Pir-Panjal Range. The vegetation of the state has undergone a drastic change. Large tracts of forest have been removed through continuous felling of trees, and bringing more area under plough and pastures. Area under forest cover district-wise in J&K State is shown in Table 12.2.

Forests have numerous productive, protective and bioaesthetic functions. On the average the forests of Jammu and Kashmir yield 617, 310 cubic million feet of timber and 120,300 cubic million fuel wood annually. Moreover, they fetch about Rs. 65 crores annually to the State revenue. According to the statistic of 1991, the total forest cover of the state was about 18,590 sq. km.

The great heterogenity in the geo-ecological, pedological and climatic conditions of the state as stated at the outset have resulted into great diversity of fauna. Over 4,000 species belonging to 1500 genera are found in the State. For a systematic study the vegetation of the Jammu and Kashmir State may be classified broadly under the following types:

Sub-Tropical Forests

The southern parts of the Jammu Division with an elevation below 1500 metres have the sub-tropical vegetation. A great variety of vegetation is found in this zone on account of edaphic factors. These forests have a thick undergrowth of bushes and scrubs. The dominant species of this zone are Shisham, Neem (*Azadirachta indica*) Pipal (*Ficus religiosa*), Tut, Kikar, Babul (*Acacia arabica*), Khair (*Acacia catechu*) thorny bushes, evergreen shrubs, climbers and tall grasses locally called as Khair. Most of these species are broad leaved deciduous type which drop their leaves in the months of January, February and March before the advent of hot weather season. Apart from fuelwood, these forests are utilized for the roofing of houses, agricultural implements and timber. Valuable products like resin, *Katha* and medicinal herbs are also obtained from these forests.

Table 12.2: J&K State–Area Under Forest–District-wise

Sl.No.	*District*	*Area in sq.km*	*Percentage*
1.	Anantnag	2289.21	11.34
2.	Badgam	429.65	2.12
3.	Baramulla	2504.83	12.41
4.	Kupwara	1479.68	7.33
5.	Pulwama	1008.76	5.00
6.	Srinagar	411.86	2.04
7.	Doda	5403.27	26.78
8.	Jammu	1445.49	7.16
9.	Kathua	789.25	3.96
10.	Poonch	825.53	4.06
11.	Rajouri	1296.40	6.42
12.	Udhampur	227.69	11.29
13.	Kargil	11.20	0.05
14.	Leh		
	Total	20173.74	100.00

Source: Forest Statistics Division, J&K (1984).

Temperate Forests

Going northward as the latitude and altitude increase the place of sub-tropical forest is taken by temperate vegetation. The slopes of Pir-Panjal and Zanskar above 1500 m. to 3000 m. are dominated by coniferous vegetation, the dominant species being deodar (*Cedrus deodara*), pine (*Pinus*), Silver-fir (*Abies alba*), Spruce (*Picea*), laural, elm and paper-birch.

The northern slopes of Pir-Panjal have the dominance of *deodar*, blue-pines with deciduous shrubs, while the Jhelum Forest Division (stretching over Gulmarg and Lolab Valley) is dominated by cedar, fir and spruce species of conifers. In the flat Valley of Kashmir, *Poplar*, *Chinar* (a type of maple), *Willow* are the main species of vegetation which are deciduous in character.

Most of the species are utilized for fuel, wood and timber. The conifers are of great economic importance. These forests produce good quality of timber and paper pulp. The most valuable timber that of *deodar* has been depleted and now confined to the north-west corner of Kashmir Division. The system of farming-out timber to contractors has resulted in the destruction of many fine forests. The tracts which the contractors have spared have suffered at the hands of the Gujjars who with small axe cut down trees partly for the sake of fuel wood, fodder and partly from a kind of natural instinct which impels him to make forest clearings for the sake of grass. Moreover, the inadequacy of hydel and thermal electricity and the severe cold in winters have put the people in position which tempt them to cut trees and their twigs. The accelerated rate of felling of trees is leading to the extinction of deodar, spruce, pine and fir. The over exploitation of forests and the use of costly wood for fuel is doing great harm to the state revenue on the one hand and the ecosystem on the other. In fact, many of the ecosystems have lost their resilience characteristics.

Alpine Pastures (Margs)

This vegetation zone lies between 3600 to 4000 m and begins well above the upper limit of conifer forests. Being the zone of high altitudes the climate in alpine pastures is extremely cold over the greater parts of the year. The low temperature supports some dwarf birch and junipers making a shrubby appearance. The stunted trees merge into vast pastures, locally known as *Margs* (alpine meadows). In these pastures some shrubs form isolated thickets during summer months (June-August). The green and nutritious grasses of alpine pastures are utilized and grazed by the Gujjars and Bakarwals who practice transhumance. These people ascend with their goats and sheep and occupy them in the month of May. They stay there upto the middle of September, depending on the prevailing temperature and precipitation conditions. In winters they descend and migrate with their flocks to the zones of lower altitudes of Jammu Division to pass their winters.

Xerophytic Vegetation of Ladakh

The high altitude desert land of Ladakh with practically no rainfall over greater parts of the year is largely dominated by xerophytic vegetation. The natural tree elements being largely lacking except for more humid region of Nubra and other side valleys of the Indus river. Hippophae-scrub is their major woody component which is used as fuel and material of roofing of houses.

Importance of Forests

The forests with their flora and fauna, play a very important role in the life of man and the development of civilization. As a saying goes "*Forests preced civilization and deserts follow them*". We are on the verge of civilization dominating forests: Therefore, conservation and protection of natural resources should be the main slogan for the coming decades. It is high time that we orient all our efforts to ensure sustainable development with sound environment for the betterment of mankind.

Forests, also forming the abode of wildlife (including both flora and fauna), account for an important land use in India forming a quarter of land use (22.8 per cent). Forests provide goods and services. The goods obtained from forests meet:

1. Domestic needs of rural based and economically poor populace.
2. Provide raw materials to wood based industries.
3. Being employment generative, form a resources for providing continuous year round and gainful employment where it is needed most.

The other important services provided by forests includes:

1. *Oxygen-recycling process* through photosynthetic activity and recharging the atmospheric oxygen.
2. *Prevention and control* of pollution by absorbing different pollutants and cleaning the atmosphere.

3. *Maintenance of water-balance* and increasing water-yield and thus, contributing to the concept.

   "Forests mean water, water means bread and bread means life."

4. *Meet recreational needs* and bio-aesthetic concepts of humanity.
5. Provide for economic returns and vital needs of food, medicines etc.

Over-exploitation of both forests and wildlife may result in spurt in goods and services, but they loose permanently their productivity and service capacity. Forests and wild-life if destroyed amounts to destruction of property, plentifulness and adverse effects on civilization and national economy. Deforestation are parallel phenomenon causing ecological. imbalance, environmental destruction flash-floods, chronic droughts and scarcity of water conditions.

Fauna of J&K State

The State of Jammu and Kashmir has great diversity in the physiography, climate and vegetation. This diversity in geo-ecological conditions provide the required habitats to varied animal kingdom. The fauna of the state, consequently, differs from latitude to latitude and altitude to altitude. In general leopard, bear, wolf, foxes, deer, *Hangul*, nilgai and wild-cats are the major wild beasts. Their population is however, decreasing owing to the over-interaction of man with nature and his encroachment in the forest and natural areas.

Musk-deer (*Kasturi*), monkeys, Langurs, black-bear, wild pig (*Bad*) have their habitats between, 2,000 and 3,500 metres. They shift their place of stay seasonally *i.e.* in winters they descend to lower heights and in summers ascend to higher altitudes. The Central Himalayas, Zanskar-Range and higher reaches, of Pir-Panjals are the abodes of these animals.

Leopard, *chetah*, wolves and foxes are found in the forests of Pir-Panjal, Liddar, Sindh, Doda, Udhampur, Riasi, Uri, Poonch, Kamah, Kupwara and Naranag catchment. Black-bear is very common in Lolab, Liddar, Doda, Rajouri and Uri tracts. *Barasingha* (Hangul) are also found in these localities.

Snow partridge, Himalayan-cock, *chikor* (Kak), cuckoo, larks, *bulbuls*, quail, storks, parrots, pigeon, dove, ducks, *Teel*, Vulture, crow, eagle and owls are the main birds of the state. A wide variety of reptiles, snakes, fish and insects are also found in the State.

Environmental Hazards

The environmental hazards in the Himalayas are many, complex, and interrelated *e.g.* landslides, droughts, floods, decline in the forest cover, decline in agricultural yield, soil loss, problem of siltation in reservoirs and rivers, extinction of rare species of wild animals, etc. The natural springs are rapidly dying, domestic animals are less productive or unproductive, the fodder resources are limited and natural forests are fast disappearing in the Himalayas.

Blasting of the rocks for the construction of roads, tunnels, dams, colonies. etc. and haphazard method of mining have greatly affected the ecological balance. Mining operations are old and traditional and have erred the lush green vegetation cover. The mining operations have also eroded the lush green hill slopes, destroyed leaf mould–a valuable organic fertilizer, polluted streams, and decreased the agricultural production by 60 per cent. In the Doon Himalayas, the limestone dust particles from the quarries and lime factories have been estimated at 360 micrograms per cubic metre as against the safety level of 100. Hazardous air and noise pollution have caused many plants to wither away; the forest under-growth has been cleared, the sub-surface water has decreased, and the mountain side resembles a gigantic skeleton due to blasting (Choudhary, 2008; Gupta *et al.*, 2001; Mahajan *et al.*, 2004 and Trivedi, 2002).

In higher altitudes, forest edging is an environmental hazard. The glaciers do a lot of abrading and scratching and sometime snow streams and avalanches cause a great loss to life and ecology. In areas of moderate relief of the Himalaya, water erosion is a serious problem as it makes the rock naked; consequently, instability of the hill slopes, silting of the rivers valleys and floods are very common features. Torrential rains cause mass wasting and huge landslides and landslips. Landslides are very common, often blocking the vehicular traffic and sometimes damaging engineering projects. In the low relief alluvial plains, flash floods, soil erosion and sheet erosion are environmental hazards.

The construction of unimaginatively planned and cruelly executed roads is the biggest factor in destroying the ecological balance. Construction of dams on hill rivers to meet the ever increasing demand of irrigation water and power of the plains is yet another environmental hazard (Figures 12.4–12.5). These projects cause destruction of hills, flood the fertile flatland of the valleys, and create the problems of rehabilitation of the oustees.

To prevent or reduce degradation of the Himalayan natural resources there should be scientific and well planned growth of the region to which local people should have close involvement in planning. There should be recreation between natural resources and tourism development which should be according to the carrying capacity of particular resorts, so that space action does not affect their ecosystems and ecological balance. If these protective measures are not given due importance in tourism planning and development, the Himalaya would be damaged beyond the limit of repair of regeneration.

Industrial development in the Kashmir Himalayas is slow but the rapid population explosion and urbanization has changed the trend. In the Kashmir Himalaya, like the Himachal Himalaya, the Kumaon Himalaya, the Assam Himalaya and in many other parts of the Himalayas fruit processing, wool processing, cottage industries, cement industry, paper and pulp industries, resin processing and other forest based industries, with pollution control devices would be vital for the rapid economic growth of the region.

Measures to Preserve the Ecology

The Himalayas thus provide a variety of natural resources to the mankind including the life giving water. These resources have been threatened largely by human negligence and activities. In order to improve and regenerate degraded hill environment, scientific approach involving hazard zone mapping and banning engineering development activities in such zones, drainage control and dewatering measures, protection of slopes through vegetation, modification of slopes where necessary, application of engineering technology and improved agronomic practices should be carried on a priority basis. The ecological balance should be maintained and improved by reforestation, checking the illegal felling of trees, controlling the course of floods by construction of bunds and

Figure 12.4: Silting of River Chenab due to Landslides, and Construction Activities at Dul-Hasti, J&K State

Figure 12.5: Construction of Unimaginatively Planned and Cruelly Executed Road Alongside of Neeru Stream at Bhaderwah, J&K State

Figures 12.4–12.5: Anthropogenic Activities Responsible for Causing Environmental Hazards in Kashmir Himalayas (Photo by Dr. A.K. Verma)

Figure 12.6: Mighty River Chenab Passing through Doda District of J&K State

Figure 12.7: Tapping and Plugging of the Chenab River Downstream for Construction of Dam to Generate Hydro-electric Power at Dul-Hasti Project

Figures 12.6–12.7: Exploitation of Natural Resources in Kashmir Himalayas (Photo by Dr. A.K. Verma)

retaining walls, checking landslides, rocketries, sinking of the soils, protecting the watersheds, etc.

Laws should be enacted to stop indiscriminate timber trade, installation of stone crushers, slate factories, industries, etc., which in turn cause air, water and noise pollution. The mining operations should be done using modern methods and techniques to minimize the pollution of atmosphere, the ruthless and haphazard mining activities greatly affect the health of the workers and also create ecological imbalance by rock debris, dumps and quarries waste. The greenery of the soils is lost and fertility of lands is decreased. Constructional and developmental works also destroy the natural setting of the scenery and landscape.

Soil and land use inventories need to be taken up on priority basis. Modern techniques of remote sensing and serial photo interpretations are very vital in conserving the soil and soil programme. Soil maps may be employed to carry out the afforestation programmes and proper land use. Aerial seeding should be carried out to reclaim the barren lands.

To prevent or reduce degradation of the Himalayan natural resources, there should be scientific and well planned growth of tourism in which experts of different disciplines like geology, botany, geography, wildlife, environment, etc. and technocrats like engineers, architect etc. along with local people should have close involvement in planning. There should be correlation between natural resources and tourism development which should be according to the carrying capacity of particular resorts, so that the space action does not affect their ecosystem and ecological balance.

If these protective measures are not given due importance in tourism planning and development, the prosperous heritage and environment of the Himalaya would be damaged beyond the repair of regeneration. To protect our environment we have to protect our wildlife, rare birds and unique plants. Proper education on environmental rehabilitation and conservation must be initiated at the grass root level. Rehabilitation and conservation programmes should include such measures as finding alternative sources of energy, self employment, agriculture, sericulture, horticulture, social forestry, cottage industries, small scale industries, etc. The human and animal husbandry programmes should also be taken up to improve the lot of the Indian village folks. Women are the backbone of human ecosystem in the Himalayan hills. It is, therefore, necessary to improve their lot. An all-out effort should be made to maintain the enchanting beauty of the Himalaya.

It is only during the last few decades that man has begun to exploit the vast and varied natural resources. The great rivers which originate from the springs and the snowfields possess immense potential for a multiplicity of uses. At several places these rivers have been tapped and plugged to generate hydroelectric power, to construct dams and generate power, to irrigate the land, and to transport the material through the rivers (Figures 12.6–12.7). A proper management of these rivers and other natural resources to benefit the inhabitants of the holy mountains is very vital. The problems of environmental management are diverse and require special attention.

Conserving and Managing the Depleting Resources

The Himalayas extend from the west to the east for about 2,500 km and they vary in width from 250 km to 300 km. These mountains have the world's highest peaks. The diverse landscape soils and climatic sectors, support several types of vegetation. Some of the plants are very characteristic of tropical, sub-tropical, cool temperate, temperate and alpine conditions. Some of the regions also support desert or semi-desert vegetation. The Himalayan ranges can be divided into three major botanical regions *i.e.,* the western, the central and the eastern Himalayas. The central Himalayas more or less, conform to the mountainous region in Nepal. The eastern Himalayas includes Sikkim, Bhutan, Arunachal Pradesh, etc.

Considering that the Himalayan region is richly endowed with a wide range of plants and is, in fact, one of the centers of origin of certain important agricultural and economic plants, an urgent need is to survey plant wealth, especially rich in genetic diversity. The survey, however, should not be merely a taxonomic one but a dynamic one carried out by multidisciplinary experts with the object of conserving the rare species. The exotic plants, like wild roses, coniferous trees, and rare herbs and food crops should be genetically upgraded to meet the requirement of food.

The Himalayan belt offers great potential for forage grasses and legumes, as such the natural grasslands of low productivity can be replaced with highly productive ones by farming systems. These untapped grassland resources of the Himalayan region can perhaps make the largest contribution to its economic development. As the prosperity of a nation is chiefly determined by its capacity to utilize and conserve its resources, it is therefore, necessary to undertake steps based on the physical and biological factors, so as to formulate a plan for the rapid development, economic progress and sound management of the Himalayan region.

Conclusion

The natural resources of the region have, over the years, been subjected to serious depletion owing to various factors, stemming mostly from anthropogenic activities. As elsewhere in the country, both land and water resources in the Himalaya are beset with environmental problems. Deforestation, overgrazing, unsuitable agricultural practices, especially on sloping lands, strip the land of its natural cover/vegetation, making the land vulnerable to soil erosion and loss of major soil nutrients. The adverse impact of such denudation and soil erosion in the Himalayan watershed/catchment areas is a matter of grave ecological concern. Management of land and water and environment in general is governed and facilitated by a set of policies/legislations that, if implemented properly and effectively, should support biodiversity conservation and sustainable utilization. However, the development of the Himalayan region depends on the enhancement of biomass productivity through sustainable management of the natural resources that ensures the conservation of species, including the genetic variability within them, and eco-systems. As such sustainable management of natural resources along with environmental conservation has to be an integral element of the development process.

References

Choudhary, A.R., 2008. Biodiversity conservation in Himalayas. *Daily Excelsior Magazine-Eco-watch,* Dated 13th May.

Gupta, A., Das, M. and Kapoor, A.N., 2001. *Principles of Physical Geography*. S. Chand and Co., Ltd., Jallandhar.

Mahajan, R., Verma, Anil K. and Manhas, Rohini, 2004. Forest ecosystem. In: *Handbook of Field-cum-Laboratory Work in*

Environmental Studies. National Book Depot (Publisher), Udhampur, J&K State, 146 pp.

Trivedi, P.R., 2002. Himalaya. In: *Environmental Education: Ecology and Environment III*. Indian Institute of Ecology and Environment Publications, New Delhi, pp. 59–75.

Chapter 13

Plant Genetic Resources of a North-Western Indian State*

N.S. Chauhan and Sanjeev Thakur

ABSTRACT

Himachal Pradesh situated in the North-Western Himalayan zone is rich in plant genetic resources. It excels in having thick populated Oaks, Conifers, Betuta and highly valuable medicinal and aromatic plants. As the demand for forest products is on the increase, overexploitation of forest wealth has taken place and is a cause of concern. Many valuable plant species are threatened and face danger of extinction. The article deals with plant genetic resources of Himachal Pradesh and the strategies needed for their conservation.

Introduction

The green forest wealth, a renewable resource, has experienced a colossal depletion in the recent past in India in general and in the Himalayas in particular. The increasing population coupled with intensive resource mobilization has led to the transfer of areas previously held under forest to agriculture and other uses (Willian, 1973). The large scale deforestation is leading to an accelerated loss of valuable or potentially valuable biodiversity, particularly in those areas where botanical or gene ecological exploration have not been carried out. Through indiscriminate exploitation, destruction of habitats, excessive use of harmful chemicals, a number of plant

* *Chapter Courtesy*: Vinod Publisher & Distributors, Jammu.

species have already disappeared while others await a similar fate, at times, even without our being aware of their existence. According to an estimate of Threatened Plants Committee of International Union for Conservation of Nature and Natural Resources (IUCN) about 10 per cent (20-30,000) of World's flowering plants are reported to be "dangerously rare" or "under threat" (Jain and Sastry, 1980). It is feared that by the year 2000 A.D., two fifth of the species and by the 2020, one third of the species will be lost, if the present trend continues (Anon, 1992).

In addition to large scale disappearance of forests, these are being brought under intensive management, which may endanger certain species and change the composition of others (Kemp et al, 1976). The population at the outer limit of the range of the species is likely to be the most affected than the population occurring in central areas. Further more, these isolated or marginal populations have developed specific resistance to drought and diseases, which may be of great potential value in breeding, improvement and afforestation programmes.

Most of the forest tree species have very wide gene base as a consequence of natural outcrossing and large population occurring in a variety of habitats. Thus in normal environment, a species is equipped for both short term adaptation and long term change, because its large gene pool allows it to sort genes as needed. Improved plantations to meet the growing demand of wood product have led to a change in gene complexes which were present in the original population. Since these populations have narrow genetic base, they are not flexible to withstand any adverse change in the environment. In Himachal Pradesh over 38 per cent area was previously under forests cover. Recent satellite imageries put it around 14 per cent. This has resulted from large scale developmental activities, population pressure, increased cattle population and exceedingly extended agricultural and horticultural activities. Commercial species are extracted year after year without caring for their protection, preservation and propagation. Nearly around 100 species are facing danger to their existance.

Why Conserve Genetic Resources?

To meet any exigency in future, there is a strong need to identify and conserve the genetic resources keeping the following aspects in view:

1. Protect, conserve and propagate the potentially valuable germplasm for the present and posterity,
2. Conserve natural variability present within the species for maintaining broad genetic base,
3. Conserve superior genotypes for evolving high yielding varieties of economic species,
4. Preserve biodiversity for maintaining stability in the ecosystem,
5. Provide areas for ecological and environmental research including base line studies consistent with human needs and demands and also to provide facilities for education and training.

Genetic Resources of Himachal Pradesh and their Conservation

In order to evolve a suitable conservation strategy, it is necessary to have knowledge of our resources for meeting diverse needs of the people.

The State of Himachal Pradesh lies between 30°22'N to 33°13'N latitude and 75°23'E longitude covering an area of 55,673 sq. km. It is a mountainous state with altitude ranging from 247-7,000 metres above mean sea level. The entire area of the State mostly comprises of tertiary rocks. The soils are young and thin with considerable depth in the Valleys. The climate is humid sub-tropical in southern low areas, temperate in the middle altitudinal areas, cold alpine and glacial in the north and north-eastern higher mountainous areas (Wadia, 1975).

The wide range of altitude, topography and climatic condition have endowed Himachal Pradesh with a rich and diversified flora. The vegetation is broadly known through the accounts of Collet (1921); Puri (1960); Champion and Seth (1960); Nair (1977); Chaudary and Wadhawa (1984); Chauhan (1984, 1987 a and b, 1988, 1989 a and b, 1990 and 1991). As a result of these studies, more than 3200 plant species of higher plants have been identified in the state flora. Out of these there are over 500 medicinal and over 150 aromatic plant species, many of which are harnessed for market and bonafide local use/internal consumption. Broadly speaking, the germplasm for the state encompassing different plant resources, have been given in tabulated form under different groups.

Table 13.1: Timber Yielding Plant Resources

Sl.No.	Scientific Name	Local/Vern./Eng. Name	Availability/Area
1.	*Abies pindrow*	Koloroi, Tosh, Silver fir	2800–3800 m
2.	*A. spectabilis*	Tosh, High level fir	2800–4000 m
3.	*Acacia nilotica*	Kikar, Babul	Upto 1200 m
4.	*A. catechu*	Khair, Katha, Cutch tree	–do–
5	*Acer caesium*	Mandar, Kaingu, Maple	1500–3000 m
6.	*A. caudatum*	Jarima, Parag, Maple	2500–33000 m
7.	*A. pictum*	Kandal, Kanzal, Maple	1500–2800 m
8.	*A. oblongum*	Pangoi, Padag, Maple	Below 1800 m
9.	*A. pentapomicum*	Teon Kinnaur	1500 m
10.	*A. illosum wall.*	Rangeon, Kainju	200–2800 m
11.	*Adina cordifolia*	Haldu	Below 1800 m
12.	*Ailanthus altisimma*	Maharukh, Titri	1000–2800 m
13.	*Albizia lebbek*	Shirish, Siris	Below 1300 m
14.	*A. odoratissima*	Kala siris, Black siris	Upto 1400 m
15.	*A. procera*	Safed siris	Upto 1200 m
16.	*A. chinensis*	Ohi, Chiuli	Upto 1500 m
17.	*Anogeissus*	Dhav, Chhal, Dhaura	Upto 1100 m
18.	*Azadirachta indica*	Neem	Upto 1000 m
19.	*Bombax ceiba*	Semul, Rel, Silk–cotton	Upto 1500 m
20.	*Buxus wallichiana*	Papri, Boxwood	1600–2800 m
21.	*Cedrus deodara*	Devdar, Cedarwood	1600–2800 m
22.	*Celtis australis*	Khirk, Nettle tree	1200–2700 m
23.	*Dalbergia sissoo*	Shisham, Tali	Below 1300 m
24.	*Eucalyptus citriodora*	Lemon scented eucalyptus	Planted at many places
25.	*E. hybrid*	Safeda, Eucluptus	–do–
26.	*E. globulus*	Blue gum	–do–
27.	*Juglans regia*	Akhrot, Walnut	1000–2800m
28.	*Mangifera indica*	Aam, Mango	Below 1500 m

Contd...

Table 13.1–Contd...

Sl.No.	*Scientific Name*	*Local/Vern./Eng. Name*	*Availability/Area*
29.	*Morus laevigata*	Shahtoot, Mulberry	1000–1500 m
30.	*Melia azedarach*	Drek, Bakain	Upto 1800 m
31.	*Mitragyna parviflora*	Korth, Kadam, Kain	Below 900 m
32.	*Olea glandulifera*	Jharinu	Upto 1300 m
33.	*Ougenia oojeinensis*	Sanan, Sandan	–do–
34.	*Pinus gerardiana*	Neoza, Chilgoza	Kinnaur
35.	*P. roxburghii*	Saral, Chir	500–2100 m
36.	*P. wallichiana*	Kail, Blue pine	1500–2800 m
37.	*Picea smithiana*	Rai, Spruce	2000–3200 m
38.	*Populus ciliata*	Biun, Himalayan polpar	Upto 1500–3200 m
39.	*P. eupharatica*	Bahn, Indian poplar	Upto 400 m
40.	*P. balsamifera*	Balsam poplar	Cold desert
41.	*Quercus glauca*	Bani	Below 1500 m
42.	*Q. leucotrichophora*	Bani	1200–1800 m
43.	*Salix alba*	Willow	Lahaul
44.	*Syzygium cumini*	Jamun	Below 100 m
45.	*S. tetrasperma*	Bhashsuli, Willow	1200–2800 m
46.	*Terminalis arjuana*	Arjun	Upto 1000 m
47.	*T. tomentosa*	Asan, Sain	–do–
48.	*Toona ciliata*	Tun	Upto 1300 m
49.	*Ulmus laevigata*	Marinu	Below 1500 m
50.	*Zizyphus mauritiana*	Ber	Below 1200 m

Table 13.2: Wild Fruit Bearing Plant Resources

Sl.No.	*Scientific Name*	*Local/Vern./Eng. Name*	*Availability/Area*
1.	*Aegle marmelos*	Bilve, Bil	Below 1200 m
2.	*Aesculus indica*	Kanor, Khanor	1800–3200 m
3.	*Artocarpus lakoocha*	Dheu, Lakuch	Below 1300 m
4.	*Berberis aristata*	Kashmoli, Chotahr	Below 1500 m
5.	*Celtis caucasica*	Karondha, Garna	Upto 2800 m
6.	*Cordian dichotma*	Lasura	Below 1200 m

Contd...

Table 13.2–Contd...

Sl.No.	Scientific Name	Local/Vern./Eng. Name	Availability/Area
7.	*Corylus colurna*	Ban shiroli, Bhutia badam	1800–3200 m
8.	*Ehretia acuminata*	Puna, Heliotrope tree	Below 1500 m
9.	*Eleaqnus umbellata*	Giyanyin, Guan	1000–2600 m
10.	*Emblica officinalis*	Aonla, Amla	Below 1200 m
11.	*Ficus auriculata*	Tryamblu, Tirmal	Upto 1800 m
12.	*F. carica*	Anjeer	–do–
13.	*Hippophae rhamnoides*	Chharma, sea buckthorn	2800–4500 m
14.	*Juglans regia*	Akhrot, Walnut	Upto 2800 m
15.	*Lonicera aungustofolia*	Piddu, Bandhakh	2000–3000 m
16.	*Madhuca latifolia*	Mahua	Below 1000 m
17.	*Mangifera indica*	Aam, Mango	Below 1200 m
18.	*Morus alba*	Shahtoot	Upto 1500 m
19.	*Phoenix humilis*	Khajuri	Upto 1300 m
20.	*Pinus gerardiana*	Chilgoza, neoza	1800–3000 m
21.	*Prunus armeniaca*	Chuli	1500–3800 m
22.	*P. persica*	Jangli aru, Bbaimi	1800–3600 m
23.	*Pyrus communis*	Nakh, Nashpati	1000–2200 m
24.	*Punica granatum*	Daru, Dharim	Below 1500 m
25.	*Ribes alpestra*	Ikuta	2800–4000 m
26.	*Rubus ellipticus*	Akhe, Hisalu including and allied species black, red purple black type	1600–2800 m
27.	*Spondias pinnata*	Ambara, Amaratka	Below 1000 m
28.	*Syzygium cumini*	Jamun, Jambu	Below 1200 m
29.	*Tamarindus indica*	Imli	Below 700 m
30.	*Terminalia bellirica*	Bahera	Upto 1200 m
31.	*T. chebula*	Harad	–do–
32.	*T. catapa*	Jangli badam	Below 1000 m
33.	*Vibrunurn cylindrica*	Tila	1200–2400 m
34.	*Zizyphus mauritiana*	Badir, ber	Upto 1300 m
35.	*Z. nummularia*	Jharberi, Bubadari	Upto 1300 m
36.	*Z. oenoplia*	Gidar, Ber, Makoh	Below 700 m
37.	*Z. sativa*	Beri, Kandiari	–do–

Table 13.3: Fodder Resources

Sl.No.	Scientific Name	Local/Vern./Eng. Name	Availability/Area
1.	*Acacia modesta*	Kikar, Babul	Upto 1500 m
2.	*Acer acuminatum*	Maple	1800–2800 m
3.	*Aesculus indica*	Kanor	1400–3200 m
4.	*Albizia chinensis*	Ohi, Chuli	Upto 1300 m
5.	*Alnus nepalensis*	Kunish	1300–2800 m
6.	*Adina cordifolia*	Haldu	Below 1000 m
7.	*Anogrissus latifolia*	Chhal, Dhav	Below 1000 m
8.	*Artocarpus lakoocha*	Dheu	Upto 1300 m
9.	*Bauhinia variegata*	Kachnar	–do–
10.	*Bamusa arundinacea*	Maggar	Upto 1200 m
11.	*B. nutana*	Nal	–do–
12.	*Celtis austrialis*	Khirk	Upto 2800 m
13.	*Coriaria nepalensis*	Gangar	1200–2400 m
14.	*Cotin coggyria*	Tung	1000–1800 m
15.	*Emblica officinalis*	Aonla	Upto 1300 m
16.	*Ficus auriculata*	Timla	Upto 1400 m
17.	*Grewia asiatica*	Phalsa	Below 1200 m
18.	*G. optiva*	Biul	Upto 1500 m
19.	*Kydia calycina*	Pulah	Below 1100 m
20.	*Morus alba*	Shahtoot	Upto 1200 m
21.	*M. indica*	Tut	Upto 1500 m
22.	*M. serata*	Kimu, Hisnu	Upto 2800 m
23.	*Phoenix humilis*	Khajuri	Below 1200 m
24.	*Populus species*	Safed poplar, Biun, Balsam	1200–3800 m
25.	*Quercus glauca*	Banni	800–1800 m
26.	*Q. floribunda*	Moru	1800–2800 m
27.	*Q. leucotrichophora*	Ban	1200–2600 m
28.	*Q. semicarpifolia*	Kharsu	2100–3800 m
29.	*Salix alba*	Biuns	1500–3000 m
30.	*S. fragilis*	Changma	Upto 4500 m
31.	*S. tetraspma*	Bhashli	Below 1200 m
32.	*Ulmus laevigata*	Marinu	Below 1500 m
33.	*U. wallichiana*	Morun	1500–2800 m
34.	*Woodfordia fructicosa*	Dhai	Below 1500 m
35.	*Zizyphus mauritiana*	Ber	Below 1200 m

In addition to these around 200 species of grasses occur which forms major forage source for browsing animals.

Table 13.4: Plant Resources Yielding Fibre and Flosses

Sl.No.	Scientific Name	Local/Vern./Eng. Name	Availability/Area
1.	*Agave america*	Ramban	Shiwalik zone
2.	*Bauhinia vahilii*	Taur	Below 1300 m
3.	*Bombax ceiba*	Semal, Silk cotton	Below 1500 m
4.	*Calotropis procera*	Ak, Madar	Kangra, Una
5.	*Cannabis sativa*	Bhang	Upto 2600 m
6.	*Corchorus capsularis*	Chanchu, Jute	Below 1000 m
7.	*Crotalaria juncea*	Kuekra	Green manure
8.	*Debregeasia hypoleucas*	Siaru	Upto 1500 m
9.	*Gerardinia hetrophylla*	Karal	1500–2500 m
10.	*Grewia optiva*	Biul	Below 1500 m
11.	*Kydia calycina*	Pullah	Below 1100 m
12.	*Linum usitatissinum*	Alsi, Linseed	Below 1300 m
13.	*Marsdenia tenacissima*	Moorva	Upto 1800 m
14.	*Typha elephantina*	Gundra, Boj	Upto 1400 m
15.	*Urena lobata*	Vilayti san	Below 1200 m
16.	*Ulmus laevigata*	Marinu	Below 1500 m
17.	*U. wallichiana*	Morun	1500–2800 m

Bee Flora Available in the State

Banana (*Muṣa paradisiaca*), *Citrus* species, wild date palm (*Phoenix sylvestris*), *Rubus* species, *Malus domestica* (Apple). *Pyrus communis* (Pear), *Prunus* species, *Mangifera indica*, *Punica granatum* (Anar), *Zizyphus* species, *Ageratum conyzoides*, *Centtaurea cyanus*, *Poinsettia pulcherrima*, *Altheaea rosea*, *Robinia pseudacacia*, *Anemone*, *Impatiens*, *Rabdosia rugosa*, *Eucalyptus*, species etc.

Oil Seed Bearing Plant Resources

Arachis hypogea, *Linum usitatissimum* (Lineseed, alsi), *Brasica campestris*, *Ricinus communis* (Castor oil), *Glycine max* (Soyabean), *Azadirachta indica* (Neem), *Aleurites fordii*, *Junglans regia* (Akhrot),

Prunus armeniaca (Chuli), *Prinsepia utilis* (Bekhal), *Sapium sebiferum* (Paras pipal), *Quercus* species (Ban, mohru, Kharshu), *Shorea robusta* (Sal), *Olea ferruginia* (Kaku, Jangli, Jaitun).

Table 13.5: Fuelwood Resources

Sl.No.	*Scientific Name*	*Local/Vern./Eng. Name*	*Availability/Area*
1.	*Acacia catechu*	Khair	Upto 1200 m
2.	*A. modesta*	Khairi	Upto 800 m
3.	*A. procera*	Safed siris	Upto 100 m
4.	*Abies pindrow*	Tosh, Fir	Above 2800 m
5.	*Albizzia spp.*	Siris, Ohi, etc	Upto 1500 m
6.	*Alnus nepalenisis*	Udis	1500 m
7.	*A. nitida*	Kunish, Kuns	1500–2500 m
8.	*Anogeissus latifolia*	Dhaura	Below 1000 m
9.	*Berberis aristata*	Kashmal and other spp.	Above 1800 m
10.	*Bombax ceiba*	Simul	Below 1300 m
11.	*Butea monosperma*	Palash	Below 1200 m
12.	*Cajanus cajan*	Arhar	Below 1200 m
13.	*Callicarpa macrophylla*	Daya	Below 1200 m
14.	*Cassia fistula*	Amaltas	Below 1000 m
15.	*Celtis australis*	Khirak	1000–2500 m
16.	*Colebrookia oppostifolia*	Binda	Upto 1300 m
17.	*Cotinus coggygria*	Tung	Upto 2000 m
18.	*Cotoneaster bacillaris*	Rheush	Above 2000 m
19.	*Dalbergia sissoo*	Talli	Below 1200 m
20.	*Debregeasia hypoluca*	Siaru	Below 1300 m
21.	*Dodonaea viscoas*	Mendru, Aliar	Below 1200 m
22.	*Eleagnus latifolica*	Giyaindji	Upto 2800 m
23.	*Engelhardtia spicata*	Silapoma	Below 1200 m
24.	*Erthrina suberosa*	Pangra	Below 1200 m
25.	*Fraxinus floribunda*	Angan	Temperature zone
26.	*Grewiaoptiva*	Biul	Below 1500 m
27.	*Hypericum oblongifoium*		1200–1800 m
28.	*Juniperus commuis*	Bethar	Sub-Alpine zone

Contd...

Table 13.3–Contd...

Sl.No.	Scientific Name	Local/Vern./Eng. Name	Availability/Area
29.	*Lantana camara*	Phulnu	Chronic shrub below 1500 m
30.	*Machilus gamblei*		Below 1500 m and often upto 2400 m as in Chhajpur
31.	*Mallotus phillipinensis*	Kamal, Kamila	Below 1500 m
32.	*Melia azedarach*	Drek, Bakain	Upto 1800 m
33.	*Mimosa rubivaulis*	Shiahkanya, Aila	Upto 1300 m
34.	*Moringa oleifera*	Soanjana	Below 1000 m
35.	*Morus aiba*	Shahtut	Upto 1500 m
36.	*M. laevigata*	Tut	Upto 1500 m
37.	*Murraya koenigii*	Gandhelu	Upto 1300 m
38.	*P. roxburgihii*	Chirr	Below 1500 m
39.	*Populus ciliata*	Biun	1400–3200 m
40.	*Quercus floribunda*	Moru	1800–2800 m
41.	*Q. gluaca*	Bani	Below 1500 m
42.	*Q. baloot*	Holly	Kinnaur
43.	*Q. leucotrichophora*	Ban oak	1200–2400 m
44.	*Rhamunus purpurea*	Pruple buckthorn	Below 1200 m
45.	*Rhus parivflora*	Tumr	Upto 2200 m
46.	*Salix fragilis*	Changma	Kinnaur
47.	*S. elegans*	Binus	2000–3500 m
48.	*Toona ciliata*	Toon	Upto 1300 m
49.	*Ulmus wallichina*	Morun	1500–3000 m
50.	*Vitex negundao*	Bana, Sura	Upto 1600 m
51.	*Woodfordia fructicosa*	Dhai, Dhataki	Below 1300 m

Tans and Dye Yielding Plants

Bark Tans

Acacia nilotica, Juglans regia, Bridelia retusa, Quercus species, and *Emblica officinalis*.

Leaf Tans

Anogeissus latifolia, Carissa spinarum, Emblica officinalis, Impatiens balsamifera, Acacia nilotica, Zizyphus xylopyrus, Rhus succedania, Elscholtzia polystachya.

Fruit and Gall Tans

Species of *Rumex, Gerananium, Polygonum, Juglans regia, Toona ciliata, Onsoma echoirdes, Arnebia benthami*

Plants being exploited commercially for trade and demanded by the drug factories.

Table 13.6: Plants Exploited Commercially

Sl.No.	Scientific Name	Local/Vern./Eng. Name	Part of the Plant Used
1.	*Aconitum hetrophyllum*	Atis, Patis	Roots
2.	*Acoros calmus*	Bach, Bare	Rhizomes
3.	*Angelica glauca*	Chora	Roots
4.	*Aegle marmelos*	Blive	Fruits, Rind, Bark
5.	*Artemesia maritima*	Nurcha, Seinski	Aerial parts
6.	*Berberis lycium*	Kashmal	Roots
7.	*Berginia ciliata*	Pashanbhed	Rhizome
8.	*B. stracheivi*	–do–	–do–
9.	*Cassia absus*	Chaksu	Seeds
10.	*Cinnamomum tamla*	Tejbatra	Leaves
11.	*Costus speciosa*	Kumuk	Rhizomes
12.	*Carum cariv*	Kalajira	Fruits
13.	*Datura stramonium*	Dhatura	Seeds, whole plant
14.	*Dioscorea deltoidea*	Kemuk	Rhizomes
15.	*Ephedra gerardiana*	Tutgghantha	Twings
16.	*Emblica officinalis*	Amla	Fruits
17.	*Heracluem candicus*	Padara	Roots
18.	*Holarrhena antidysentrica*	Kutaj, Kura	Bark
19.	*Helicteres isora*	Marorphli	Twisted fruits, bark
20.	*Jurinea dolmieaea*	Dhoop	Roots
21.	*Dactylorhiza hatagirea*	Salampanja	Roots

Contd...

Table 13.6–Contd...

Sl.No.	Scientific Name	Local/Vern./Eng. Name	Part of the Plant Used
22.	*Juniperus macropoda*	Dhupi, Hauber	Fruit
23.	*Physochlaina praealta*	Laltang	Leafy Plants
24.	*Picrorhiza kurroa*	Karu, Kutki	Rhizomes
25.	*Podophylum hexandrum*	Bankakri	Roots
26.	*Rheum australe*	Revanchini, Rhubarb	Roots
27.	*Saussurea lappa*	Kuth	Roots
28.	*Terminalia belerica*	Bahera	Fruits
29.	*T. chebula*	Harad	Fruits
30.	*Thalictrum foliosum*	Pilijari, mamiri	Roots
31.	*Valeriana jatamansi*	Tagar, Nihani	Roots
32.	*Viola serpens*	Banaksha	Fruits, Flowers with leaves
33.	*Morchella esculanta*	Guchhhi	Fruiting body

List of Threatened/Endangered Plant Species in H.P.

1. *Acer caesium*
2. *Aconitum deinorrhizum*
3. *A. falconeri*
4. *A. heterophyllum*
5. *A. leavei*
6. *Agropyron duthiei*
7. *Amphicarpaea edgeworthii*
8. *Aquilegia pubiflora*
9. *Artemisia amygdalina*
10. *Asphodelus commosus*
11. *Astragalus leucocephalus*
12. *Astraglus rhizanthus*
13. *Atropa acminata*
14. *Balanophora diocia*
15. *Berberis pseudoumbellata*
16. *B. royleana*

17. *Calamogrostis garhwalensis*
18. *Campanula nathaua*
19. *Caragan gerardiana*
20. *Carex borii*
21. *C. munroi*
22. *Cautleya petiolata*
23. *Colchicum luteum*
24. *Cypripedium cordigerum*
25. *C. plantaginea*
26. *C. puberula*
27. *Dactylorhiza hatagirea*
28. *Delphinuium koelizii*
29. *D. roylei*
30. *Deutzia amurenis*
31. *Deyeuxia kashmeriana*
32. *Dioscorea deltoidea*
33. *Drosera burmanni*
34. *Elsholtzia densa*
35. *Eria alba*
36. *Elymus thomsonii*
37. *Eulophia hormusjii*
38. *Euphrasia foliosa*
39. *E. microcarpa*
40. *Festuca lucida*
41. *Eramurus himalaicus*
42. *Galium confertum*
43. *Gastrodia orobanchoides*
44. *Hedysarum astragaloides*
45. *H. microcalyx*
46. *Hemipilia cordifolia*
47. *Habenaria arculata*
48. *Indigofera simlensis*

49. *Iris milesii*
50. *Lilium polyphyllum*
51. *Nardostachys grandiflora*
52. *Neottia lesteroides*
53. *Osbckia stellata*
54. *Oreorchis indica*
55. *Picrorhiza kurroa*
56. *Pinus gerardiana*
57. *Pleione humilis*
58. *Podophyllum hexandrum*
59. *Poa falconeri*
60. *P. setulosa*
61. *Polygonatum graminifolium*
62. *Primula macrophylla*
63. *Pulsatilla wallichna*
64. *Rhododendron lepidotum*
65. *Rubia edgeworthii*
66. *Saussurea bracteata*
67. *S. gnaphaloides*
68. *S. costus*
69. *S. atkinsoni*
70. *Schisandra grandiflora*
71. *Selaginella adunca*
72. *Silene kunawarensis*
73. *Thalictrum penduncullatum*
74. *Thermopsis inflata*
75. *Thylacospermum rupifrarum*

Measures for Conservation of Forest Genetic Germplasm

Conservation strategy as proposed by IUCN says, "conservation like development is for the people; while development aims to achieve human goals largely through the use of biosphere, conservation aims to achieve them by ensuring that such use is sustainable" (Given 1983). Conservation is a strategy which includes identification, collection, storage and preservation, sustained availability, restoration and regeneration of genetic resources.

Owing to persistent collection/ harvesting of commercial and useful species for trade and domestic use, many valuable resources are getting depleted at enormously fast trade. In order to regulate the extraction of these valuable resources and to conserve them for sustained use by the posterity, the measures which need to be adopted to conserve the biological diversity and to have required gene pool for use by the scientific community for conducting various studies and for bringing them under domestication leading to commercial cultivation by the farmers are discussed under different sub-heads. Based on Burley and Styles (1976) the main strategies for conservation have been outlined as follows:

Conservation of Ecosystem

Conservation of carefully selected areas of adequate size with suitable management policies would preserve not only the forest trees but also the elements of the ecosystems *viz.* plants, animals, mammals, birds, etc. This aspect can be accomplished by creating Biosphere Reserves, National Parks and Sanctuaries. Fortunately in H.P. Great Himalayan National Park and Pin Valley National Park have been created in addition to over two dozen sanctuaries.

Preservation of Rare Species and Species Threatened with Extinction

Biosphere reserves have to be created for conservation of rare and threatened species. Such reserves can be for large number of species or for a single species. *Pinus gerardiana* is one such species which is endemic in Kinnaur and Holi range in Bharmour of Chamba district. Similar reserves can be notified for conservation of rare species like *Aconite, Angelica, Junipers Rhododendrons, Saussureas, Primulas, Swertias, Arnebias* and *Colchium* in their natural habitat.

Prevention of Genetic Erosion

Broad spectrum of genetic variability should be conserved to act as a reserve for the present and future needs. Such material can be conserved *in situ* reserves or can be sampled by means of seed, pollen or vegetative material for *ex situ* conservation.

In situ Conservation

It may be achieved by adopting the following measures:

Natural Preservation

These may be wide areas, parks or natural habitats of species. Wide areas serve the purpose of conserving genes of long lived species, but not of short successional types. The most significant sites are the natural areas. According to the Society of American Foresters (1971), natural area is defined as a "tract set aside to exemplify typical or unique forest vegetation and its associated abiotic, edaphic, geological and aquatic features in as near natural conditions as possible, primarily for the purpose of science and education." In Himachal Pradesh many biosphere reserves need to be created keeping in view the available biodiversity and endangered nature of several valuable species. Apart from conservation, these provide permanent field laboratories for conducting scientific experiments. Though, there are many sanctuaries notified by the government, yet there is no protection to plant resources provided in these. Recently Govt. of India has notified the Himalayan National Park in Kullu district and Pin Valley in Spiti, which can serve as a natural reserve for conserving biodiversity of Himalayan region by imposing strict legislative and executive measures.

Standard Stands

These have been developed as seed production areas. Here, the trees are managed primarily for better seed production. The stands are generally distributed over the entire range of the species to sample variation. The trees are of average quality and their number is large enough to maintain breeding population. These can also serve as long-term preserve of germplasm. The species wise detail of seed production stands in Himachal Pradesh is given in the Table 13.7.

Table 13.7

Sl.No.	Species of Stands	Total No. (ha.)	Area of Trees	Total No.
1.	*Cedrus deodara*	33	1379.94	20,647
2.	*Pinus wallichiana*	15	968.68	5,223
3.	*Pinus roxburghii*	14	2834.20	12,562

Seed Production Areas

These are stands heavily thinned to leave best pheno-types. Size varies from 3-4 ha. They are provided with a hundered meter buffer zone to prevent outside pollination. Although many trees are removed, yet they do serve as a long term genetic reserve.

Individual Trees

Superior trees (plus or elite trees) that are used in tree improvement programmes, may be preserved in their natural state. Though, it is difficult to maintain them in nature due to natural calamities or biotic influences, yet they should be protected to test their worth. In *Shorea robusta, Grewia optiva, Pinus roxburgihii and Pinus* seed orchards are being raised (Anon, 1992). Thirty two high yielders in chir pine have also been selected (Sharma, 1986)

Ex situ Measures

Such measures refer to conservation of species outside their natural abode and can be achieved by the following methods:

Breeding Seed Orchards

Breeding seed orchards can serve as germplasm banks where collection of good clones/progenies have been maintained. In Himachal Pradesh, seedling seed orchards of *Grewia optiva* and *Pinus* spp. *roxburghii* have been established (Anon, 1991b).

Clone Banks or Seedling Banks

These may be representative of various provenances, ecotypes or special populations raised clonally or by seed.

Arborata/Botanical Gardens/Herbal Gardens

These are collection of plant materials representatives of the native flora brought together in a single place. They are mainly laid for education and scientific purpose, but serve as a germplasm reserve and provide variety of information by serving as living laboratories.

The State of Himachal Pradesh under the aegis of University of Horticulture and Forestry has undertaken the task of developing/ establishing Arboreta/Botanical gardens/Herbal gardens at different agroclimatic zones of the State. Presently such activities are in progress at Main campus, Solan, Dhaulakuan, Jach, Bajoura, Manali, Rahala, Sharbo and Tabo (Figure 13.1). In the Botanical garden at the main campus, 300 species belonging to 75 families

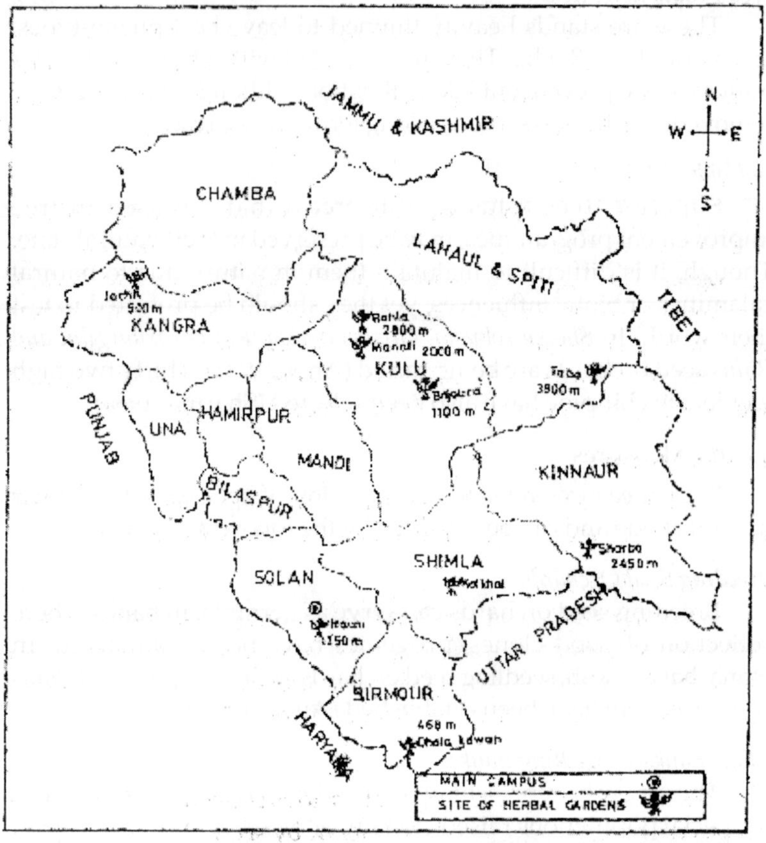

Figure 13.1: Establishment of Herbal Gardens in H.P. in Different Regional Research Stations of Dr. Y.S. Parmar University of Horticulture & Forestry, Nauni (Solan) 173 230

have been planted, where as Dhaulakuan arboratum has 64 species. Nearly 200 species of medicinal and aromatic plants have been raised in herbal gardens (Anon,1991 a).

The details of the herbal garderns is given in Table 13.8.

Ex situ Measures-Plant Parts

These include seed banks, pollen banks and tissue banks.

Seed Banks

Seed banks are the most common source of preserving germplasm. They are inexpensive and take less space. Bulked collections, provenances, single tree collections, hybrids and progeny test materials can be stored in the form of seed. In case of trees which have a long life, seed storage is lee critical. Seeds of hardwoods and that of some conifers are short lived and in such species, preserving germplasm by seed becomes difficult (Wong,1973). However, seed banks can be of a great significance for short term use in tree species and for frequent use in commercial medicinal and aromatic plant species. Seed banks of the species of *Aconitum, Ephedra, Hippophae, Arnebia, Berberis, Gentiana, Podophyllum, Rauwolfia, Polygonaturm, Valeriana* etc. can be created.

Table 13.8: Details of the Herbal Gardens

Sl.No.	*Location*	*Elevation (m)*	*No. of Species Raised*
1.	Nauni (Solan)	1250	103
2.	Bajoura	1100	36
3.	Rahla	2800	34
4.	Manali	2000	27
5.	Tabo	3800	19
6.	Jach	500	31
7.	Dhaulakuan	468	21

Pollen Banks

Pollen at low temperature and humidity can be stored for many years. Due to lack of detailed research and facilities, this method of conserving germplasm is not being used.

Tissue Banks

Short term storage of scions and cuttings is widely practiced by foresters. In recent years, much progress has been made in developing methods to preserve tissues like buds cells, callus etc. *in vitro* (Dodd, 1985). Tissue culture systems have high multiplication rate and are aseptic (free from diseases and pests). In an ideal tissue culture storage system, genetic erosion is reduced to zero. Methods that can be used *in vitro* conservation include, normal growth storage, minimal growth storage and freeze preservations.

Conclusion

Himachal Pradesh situated in the West Himalayan zone is very rich in plant genetic resources. It excels in having thick populated

Oaks, Conifers, Betula and highly commercial medicinal and aromatic plants. The demand for forest products is on the increase both for internal consumption and external demands. As a result the forests have been overexploited, and ruthlessly assaulted for a variety of reasons including developmental activities like hydel projects, roads, industrial establishments and spreading habitations. The production capacity of our forests has, as a result, reached at the lowest ebb. Many valuable species have disappeared even without our knowing them and quite a large number are facing threat to their existence. It is therefore, compelling obligation on the part of the society that the forest wealth of the state is protected, replenished and properly managed. The impact of Chipko Movement in some isolated pockets has also been witnessed. Some objectives have of course been achieved by imposing blanket ban on green felling, but in the absence of appropriate legislation and political will on the part of the government and people's active participation to impose restriction on ruthless exploitation of endangered species and regulation of extraction of minor forest products, many valuable species are getting depleted in their abode. A work force of trained and skilled manpower to achieve these objectives has to be created coupled with launching of public awareness programmes to inculcate the conservation consciousness among the masses. The elements of environment conservation should necessarily form a component of the Schools, Colleges and Universities curricula.

References

Anon, 1991a. *Annual Report*. Department of Forest Products, University of Horticulture and Forestry, Solan.

Anon, 1991b. *Annual report*. Department of Tree Improvement, Universty of Horticulture and Forestry, Solan.

Anon, 1992. *Indian Express Magazine*, June 7.

Burley, J. and Styles, B.T., 1976. *Tropical Trees: Variation, Breeding and Conservation*. Linnean Society, Oxford.

Chowdhary, H.S. and Wadhwa, B.M., 1984. *Flora of Himachal Pradesh*. BSI Publication.

Chauhan, N.S., 1984. Medicinal wealth of Pabbar Valley in Himachal Pradesh. *MSc. Thesis*, HRKVV, Palampur.

Chauhan, N.S., 1987a. *Ban Sandesh*, 20(4): 18–27.

Chauhan, N.S., 1987b. *Audhayaniki and Vaniki*, 2(12): 9–19.

Chauhan, N.S., 1988. *Indigenous Medicinal Plants*, (Ed.) P. Kaushik. Today and Tomorrow Publishers, New Delhi.

Chauhan, N.S., 1989a. *Indian Perfumer*, 33(2): 118–122.

Chauhan, N.S., 1989b. *3rd Forestry Conference*, 2: 89.

Chauhan, N.S., 1990. *J. Orchid Society of India*, 4(1,2): 99–105.

Chauhan, N.S., 1991. *Plant Resources of HP*. ST&E Council, Shimla, H.P.

Collet, H., 1921. *Flora Simelensis*. Thacker Spinks, Calcutta.

Champion, H.G. and Seth, 1960. *General Silviculture for India*. Government of India.

Dodd, J.H., 1985. *Experiments in Plant Tissue Culture*. Cambridge University Press.

FAO, 1985. FAO Forestry Paper, 20, Rome

Given, D.R., 1983. In: *Conservation of Tropical Plant Resources*, (Eds.) S.K. Jain and K.L Mehra. Department of Environment, Howrah, West Bengal.

Jain, S.K. and Sastry, A.R.K., 1980. *Threatened Plants of India*. Department of Science and Technology Government of India.

Kemp, R.H., Roche, L. and Willian, R.L, 1976. In: *Tropical Trees: Variation, Breeding and Conservation*, (Ed). J. Burley and B.T. Styles. Linnean Society, London, UK.

Nair, N.C., 1977. *Flora of Bushar Himalaya*. International Bioscience Publishers, Madras.

Puri, G.S., 1960. *Indian Forest Ecology*. Oxford Press, New Delhi.

Rau, M.A., 1975. *Illustrations of Western Himalayan Flowering Plants*. BSI, Calcutta.

Society of American Foresters, 1971. *J. Forestry*, 69: 310–361.

Sharma, K.C., 1986. In: *Proceedings of National Seminar on Tree Seed*, Hyderabad.

Wadia, D.N., 1975. *Geology of India*. Tata Mcgraw Hill, New Delhi.

Willian, R.L, 1973. *Span*, 16(3): 119–122.

Willian, R.L. and Palmberg, C, 1979. *Report on the FAO/Danida Training Course on Forest Tree Improvement*. FAO, Rome.

Chapter 14
The Present Status of Forests in the Central Himalayas*

B.L. Teli

ABSTRACT

The Indian Himalayas cover 594,427 sq.km area from Kashmir to Arunachal Pradesh. The land use figures show 53.2 per cent of area under forests. This figure gives an impression of over estimation and is disputed by many experts. The ground reality is that lately there has been unprecedented and unwise exploitation of the Himalayas by man combined with factors like immature geology, heavy rainfall, river action and seismicity resulting in landslides and other mass movements. The study deals with the present state of forests in the Indian Himalayas in general and the Central Himalayas in particular.

Introduction

Nature, by itself clothes the land with vegetation, *viz.* trees, shrubs, grasses and micro flora and fauna in varying compositions like height, density and form, which are complementary to each other. Forest management practices are largely oriented to the conservation of soil, water, vegetation and environmental purification. Forests are said to constitute the lungs of a country but in case of hills they represent the very soul of the people living there. Among the natural resources, forests are the most important ones in the hills. Forest is not only a resource but it is resource of and for other resources: Man-culture-environment as well. Other resources

---

* *Chapter Courtesy*: Vinod Publisher & Distributors, Jammu.

prosper, flourish and move on the path of development, if the natural vegetation is rich in the form, productive, economic and protective. They form an ecosystem depending on the natural environment and constitute renewable and inexhaustible resources.

The forest resources have two main functions namely protection and production. The protectional forests constitute a fragile ecosystem prone to undesirable consequences at the slightest disturbance due to geological, hydrological, soil, altitude and slope conditions. This type of forests should be taken out entirely from the orbit of the production forests and maintained only for protection of the eco-system and stability of environment in the hills. The productional function of forests include–supply of timber and bamboo, furniture, railway sleepers, defence equipments, sports goods, raw materials for paper and other industries, besides the minor products. The most vital function of forests is their role in influencing the environment, moderating the climate, maintaining the soil cover, purifying the air, conserving and regulating the water for various needs of man.

Man has to derive from nature's inexhaustible renewable resources-recreational, aesthetic, ethical, ecological and commercial and scientific values. Trees are natural and perennial land marks and sacred memorials. Some are famous and have special religious, aesthetic and sentimental associations. Man plants trees for posterity. And, they ultimately contribute to the prosperity of a nation. They are also sources of history and beauty, gifts of god and friends of man. When trees grow collectively, a forest is born, an ecological environment is created and plant and animal succession follows.

Trees interrupt reflected light, screen out undesirable sights, border the cities, lakes and streams. Trees with the rough, thick and sticky leaf surface collect soot and oil and reduce environmental pollution. Trees adorn our homesteads and our cities and shelter the wildlife. They are greatest benefactors, friends, teachers and preceptors. They have the sense of beauty and instincts of love, service, sacrifice and harmony.

The vital impacts of forests on climate, soil erosion and flood are acceptable to all. Even then extensive areas supporting natural tree growth have come under reckless exploitation and eventual intense environmental degradation. Deforestation of five million hectares of land added by ravine affected seven million hectares and another 48 million hectares lost in river valley projects offer a gloomy picture of tree cover. The claim that India has 22.2 per cent of its geographical area under forest is far from true. India has hardly 6.8 per cent of its geographical area under somewhat good vegetation cover and more than 13.4 per cent area is under poor cover. Hence, the situation has turned worse owing to uneven distribution of forests. Vast erodable areas in the hilly tracts of Western Ghats, Himalayas and Central India have been denuded beyond immediate recuperation. All these leave one to think that the tree vegetation in India is now in bad shape to offer any viable safeguard against environmental hazards.

However, the department of space reveals, what the facts are. The interpretation of landsat imagery by National Remote Sensing Agency (NRSA) obtained in 1972-75 and 1980-82, shows that we have been losing good forests at the rate of 1.5 m.h. per year and hence waste land is increasing.

Table 14.1: State of Forest Cover in India

Density per cent	*Area (m.h.) 1972–75*	*Area (m.h.) 1980–82*	*Annual Rate of Loss (m.h.)*
Good forests above 30	46.42	36.02	–1.73
Poor forests 10–30	8.75	10.06	+0.22
Total	55.17	46.08	–1.51

Due to the lack of forestal cover India losses 5,334,000,000 tons soil by erosion every year which averages about 16.38 ton/ha. Permissible value for soil in U.S.A is 4.5-11.5 ton/ha. The rivers carry approximately 38.5 per cent, at the rate of 6.26 ton/ha., nearly 29 per cent of the total eroded soil is permanently lost to the sea, 10 per cent is deposited in reservoirs resulting in loss of storage capacity of 1 per cent–2 per cent every year, and 61 per cent of the soil is dislocated and transported from one place to another. Due to the soil erosion, India is losing 12,000 crore rupees every year. The soil loss is estimated to be very high in the Himalayas *i.e.*, 28.2 ton/ha. on the average every year.

The Himalayas

The Himalayan mountain range extends in a broad arc from Pakistan through India, Nepal, Bhutan and China. The elevation ranges from 300 m in the plains at the base of the mountains to peaks ranging over 8000m above mean sea level. The Himalayas are sensitive in nature, youngest in age with a unique geographical identity, tallest and most complex of the world's mountain ranges. The Himalayas can be divided into four longitudinal and distinct litho-tectonic units, each characterized by its own typical litho-tectonic features. The lower foothills, described as sub- Himalaya are represented by Siwalik Hills composed of weakly consolidated tertiary sediments. The second zone is the middle Himalaya which ranges between 600 m. and 3000 m. It is composed of weak sedimentary and metamorphic formations including shale, sandstone, limestone, gneiss, schist, marble and some granite. Soils are generally coarse-textured and shallow and highly subject to erosion if disturbed. The third zone is the great Himalaya. It includes the highest peaks and most of the land above 5000 m and snow-clad. This region is subject to extreme weathering because of the harsh climate. This region is nonvegetated and nonpopulated. Fourth one is the Tethys or Tibetan Himalayas.

The mother earth gave birth to Himalayas, some 60-70 million years ago, in the wake of the Cainozoic Era. It is spread well over 2400 km of huge crescent in India, having on an average width of 320 km. The Himalayas are a great climatic divide that exert a dominating influence on the meteorological conditions, life, culture and history of the Indian sub-continent. The Himalayas in India cover 594427 sq. km. area from Kashmir to Arunachal Pradesh with a population of 4,29,44,663 having a density of 75 per persons sq. km. (1981). These figures account for more than 16 per cent of land area of the nation and 6.3 per cent of the total population of India. The growth of population during 1971-81 was (31.4 per cent) higher than the national average (24.7 per cent). The land use figures show 53.2 per cent of area, under forest, though there are large variations, from 25 per cent in Assam to 91.3 per cent in Arunachal Pradesh These figures give an impression of over estimation. Though, the areas classified as community, civil and soyam forests under protection circle do not serve the purpose of protection or production except for meeting the subsistence requirements of fuel and fodder for the local people. The net area sown hardly exceeds 9.7 per cent and 28.3 per cent area is classified as pasture land of the total reporting area. The terrain being rugged and steep slope, there is

Table 14.2: Land Use Pattern in the Himalayas (Data in per cent)

State	*Forest*	*Non-Agri. Uses*	*Barren and Unculti.*	*Pasture Uses*	*Misc. Uses*	*Cul-Waste*	*Old Fallow*	*Curr Fallow*	*N.S.A.*	Intensity
Arunachal Pradesh	91.3	00.6	00.0	00.3	00.3	02.6	02.1	00.9	02.0	–
Assam	25.0	11.3	19.6	02.4	03.2	01.7	01.4	01.2	33.8	129.8
Himachal Pradesh	27.0	05.2	04.7	33.0	01.3	07.5	00.3	01.3	19.2	165.4
J & K	62.4	07.1	04.9	02.6	02.2	03.1	00.2	02.0	5.3	136.2
Manipur	27.2	01.2	64.2	00.0	01.1	00.0	00.0	00.0	06.3	156.4
Meghalaya	36.1	03.7	10.3	00.8	06.5	20.2	11.6	02.3	08.6	115.5
Mizoram	60.8	26.8	00.0	00.3	00.4	03.5	02.4	03.1	03.1	109.2
Nagaland	26.6	04.7	00.0	00.0	04.1	04.5	39.0	07.2	13.7	106.8
Sikkim	36.4	06.8	28.5	14.2	00.6	01.5	00.6	00.0	12.0	105.8
Tripura	55.1	00.0	11.4	00.0	09.3	00.2	00.2	23.5	152.4	–
U.P. Hills	64.6	02.2	05.4	04.1	03.5	05.9	00.6	00.4	13.3	163.0
India	22.2	06.4	06.6	03.9	01.2	05.5	03.2	04.9	46.1	123.6

Source: Indian Agriculture in Brief (1985).

always a demand for agricultural land, though, the unit cost of development is very high. Furthermore, traditional agricultural practices, high cost of inputs, unrestricted use of grazing lands, increased deforestation activities, poor socio-economic conditions and increased defence, tourism and development activities have caused an alarming situation as far as the ecological balance is concerned. The unprecedented and unwise exploitation of Himalayas by man combined with its immature geology, heavy rainfall, river action and seismicity produce a variety and class of landslides and other mass movements without any parallel in the world.

Wide range of altitudes, temperatures, precipitation and topography results in a varied and rich flora, ranging from northern tropical dry mixed deciduous forests to alpine pastures. The sub-tropical Pine forest covers the maximum area (37400 sq. km.) while the sub-tropical dry ever green forests cover only 1730 sq. km. Area.

Broadly, the Himalayan forest resources are classified in three groups:

1. Tropical and sub-tropical
2. Temperate and
3. Alpine, which cover 39.7 per cent, 43.3 per cent and 17 per cent area respectively.

Table 14.3: Distribution of Forest Types in Himalayas

Sl.No.	*Forests Types*	*Area (sq. km.)*	*% of Total Forest Area*
1.	Sub-tropical hill forest broad-leaved	2870	2.7
2.	Sub-tropical Pine forest	37400	35.4
3.	Sub-tropical dry evergreen forest	1730	1.6
4.	Montane wet temperate	16130	15.4
5.	Himalayan dry temperate	2270	2.1
6.	Himalayan moist	275250	25.8
7.	Sub-alpine/alpine forest scurb	18000	17.0
	Total	105650	100.0

The distribution of forest types in the Himalayas vary from one state to another. The sub-tropical Pine forest is found in all the Himalayan states while temperate forests are located in Western Himalayas. The conifers and broadleaved forests in the Himalayan ranges occur in moist and dry temperate regions.

Table 14.4: State-Wise Distribution of Forest Types in the Himalayas (sq. km.)

State	*Sub-tropical Pine Forest*	*Temp-erate Forest*	*Dry Temp. Forest*	*Coniferous etc.*	*Other Hill Forest*
Assam + Meghalaya	680	–	–	50	–
Haryana	250	–	–	20	–
Himachal Pradesh	7290	7530	690	7850	13800
J & K	2810	9170	1410	20860	180
Nagaland	870	–	–	260	2630
Punjab	170	–	–	130	–
Uttar Pradesh	5110	10550	170	12530	2000
Manipur	1350	–	–	–	–
Arunachal Pradesh	18890	–	–	4200	47340
Sikkim	–	–	–	240	2580
West Bengal	–	–	–	90	1340

A number of forest species are found in the Himalayas, a few important among them are given here.

Table 14.5: Area Under Different Species in Western Himalayas (sq. km.)

Species	*J&K*	*H.P.*	*U.P.*	*Total*
Cedrus deodara	1130	700	190	2020
Pinus wallichiana	1800	860	160	2820
P. roxburghii	1580	1150	4120	6850
Abies and *Picea*	2980	1480	840	5300
Quercus incana	–	560		
Q. dilatata	1020	20		
Q. semicarpifolia	–	570		

The Central Himalayas

The planning commission (1982) Govt. of India has divided the Himalayas into the eastern states of Sikkim, Meghalaya, Manipur, Tripura, Mizoram, Nagaland, Arunachal Pradesh, hill areas of Assam and Darjeeling districts of West Bengal State, the Western Himalayas consisting of the States of Jammu and Kashmir and Himachal Pradesh and the Central Himalayas covering eight hill districts of Uttarakhand.

The central Himalayas lie between 29° 5′–31° 25′ N latitude and 77° 45′–41° 0′E longitudes covering 51130 sq. km. area. It consists of eight hill districts–Dehradun, Uttarkashi, Tehri, Garhwal and Chamoli of Garhwal region and Almora, Pithoragarh and Nainital of Kumaon region. Centrally located in the Himalayas and the northern hilly part of erstwhile Uttar Pradesh State, it is also known as the Uttarakhand.

The region is characterised by highly folded, faulted and trusted rocks which slowly disintegrate giving rise to debris accumulation on slopes. It loses much of its strength by rainfall, thus producing landslides involving large mass of soils causing blockade of roads and rivers. Sedimentary rocks are intensely folded and faulted. Though the landslides are frequently associated with metamorphic rocks, shales associated with limestone are very friable. The climate ranges from sub-tropical to alpine depending upon altitude and physical features. The lower and outer hills have climate similar to plains while higher altitudes are covered with perpetual snow. The climate is mainly decided by geographical position, elevation, direction of slope and the amount of rainfall. On an average region receives 171 cm rainfall in 70 rainy days.

The region has 4.84 million populations (1981), out of which 41 per cent are workers. 85 per cent of the working population is engaged in primary activities. The density of population is 95 persons per sq. km. only 13.2 per cent area is under cultivation having low productivity and less efficient agriculture. The land-man ratio is 0.15 ha. Land holdings are small (0.99 ha.) and fragmented.

Historical Backdrop

Before the advent of Britishers, there was no well planned management of the forests in the region. The population was sparse and the forest area was very extensive. The local people could easily satisfy their needs from the forests. The forests within the boundary of a village were considered to be the common property of all the inhabitants of the village. Villagers in these forests had complete freedom of grazing cattle, lopping, collecting timber and fuel wood for their daily requirements. It was an independent territory ruled in small kingdoms till late as the end of 18th century. The region was subjugated by the Gorkhas in 1794.

The British administration in the region commenced in 1815. From 1817 to 1823, Trail made seven land settlements, which are known as SANASSI settlement and the boundaries demarcated there under are called SALASSI. Though the Salassi boundaries had no legal significance but were important because within these boundaries the villagers enjoyed rights of grazing, timber and fuel wood.

The first forest department was constituted in 1862. In 1868 the first Indian Forest Act was made, which was repealed in 1878, when new Act was made. The forest boundaries were demarcated in the region and declared as reserved forests. At present we call them 'Old Reserves'. From 1878 to 1893, the boundaries of several forest tracts in the region were demarcated and were declared as protected forests. Under a proclamation of October 17, 1893, all the forests and wasteland of the region were declared protected forests. The village people were prohibited from cutting trees within 100ft, on either side of the road and outside five miles of the boundaries of village. Restrictions were also imposed on breaking up cultivable wasteland for cultivation and hunting etc. in forests. Under another order, Deodar, Cypress, Chir, Papri, Sal, Sisam, Tun and Khair were declared as protected trees.

Following the report of Dr. Voelcker on the "Improvement of Indian Agriculture" in 1893, the Govt. of India declared its forest policy on October 19, 1895. This forest policy had definite bias for serving the agricultural interests. Forestry was treated as a handmaid of agriculture. The policy had two main objects

1. Management of forests is to promote the general well being of the country.
2. Forest is maintained primarily for the preservation of the climate and physical conditions of the country and secondarily to fulfil the needs of the people.

The above forest policy was revised in 1952 as the National Forest Policy of independent India. To fulfil the national needs the policy has advocated classification of forest into protected forests, national, village forests and tree lands.

From 1895 to 1952, when the new National Forest Policy was announced, forest administration in India was governed by this policy. It was further amended in 1980. During this period a number of forest grievances committees were set up to settle the matter between government and villagers, regarding their rights.

The data shows that the 48.53 per cent area was under forests of the total area of the region in 1950-51, and it was 65.45 per cent in 1985-86. The forested area was 49.22 per cent in 1957-58 which reduced to 25 per cent in 1959-60; More area under forest was taken in 1968-88, which is calculated to be 60.67 per cent of the U.P Himalayas. It was 62.94 per cent in 1969-70, it was 24407 sq. km. (65.45 per cent) in 1985-86, which is more than any other Himalayan State (excluding Arunachal Pradesh).

Forest Distribution

The U.P. Himalayas have 65.45 per cent area (29042 sq.km.) under forest, out of which 68 per cent is reserved forest (Forest Deptt.), 22 per cent Civil Soyam Forest, 8 per cent Panchayat Forest and Private, Cantonment and Municipal forest has 2 per cent area. The maximum percentage of forest area is found in Chamoli (18.14 per cent) and minimum in Dehradun district (7.24 per cent). It is followed by the districts of Garhwal, Nainital, Almora, Uttarkashi, Pithoragarh and Tehri having 16.03 per cent, 13.72 per cent, 13.57 per cent, 12.27 per cent, 11.38 per cent and 7.06 per cent respectively. The percentage area under forest to the geographical area of the districts varies from 88.6 per cent in Uttarkashi to 37.3 per cent in Pithoragarh. The other districts namely Garhwal, Tehri, Almora, Dehradun, Nainital, and Chamoli have 85.6 per cent, 80.6 per cent, 73.2 per cent, 71.8 per cent, 58.7 per cent and 57.7 per cent forest area respectively (Table 14.6).

Forest Types

Forest types can be defined as a vegetation unit which possess characteristics in physiognomy and structures sufficiently pronounced to permit its differentiation from other such units. A

Table 14.6: The U.P Himalayas–District Wise Forest Area (sq. km.)

District	Reserve Forest	Civil and Soyam	Panch-ayat	Private and Municipal	Total	% to Dist. Area	% of Regional Forest
Almora	1490.4	1821	628.0	0.5	3940.9	73.2	13.57
Pithoragarh	1381.0	1210	714.4	–	3305.7	37.3	11.38
Nainital	3563.6	192	207.7	21.1	3984.4	58.7	13.72
Garhwal	2469.2	1806	376.0	6.1	4657.3	85.6	16.03
Chamoli	3704.9	1043	521.5	–	5269.3	57.7	18.14
Uttarkashi	2694.3	870	–	–	3564.7	88.6	12.27
Tehri	1495.7	516	–	205.0	2216.6	80.6	7.60
Dehradun	1954.9	148	–	–	2102.9	71.8	7.24
Total	18744.0	7606	2447.6	232.6	29041.9	69.2	100.00

Source: Forest Statistics U.P, Lucknow, p. 27.

number of forests are found in the region in various composition under the wide range of conditions and locality factors-topography, climate, geology, soils and biological factors. Some of the species are adversely affected in the region due to the repeated fire, uncontrolled burning, overgrazing, over exploitation and artificial plantation. However, number of forest types are there in the region which are broadly classed as under:

1. Tropical moist deciduous forests are found on the southern outer edge. These are mixed forests having Sal (*Shorea robusta*), Khair (*Acacea catechu*), Shisham (*Dalbergia sissoo*), Haldu (*Adina cordifolia*), Semal, Tun, Bamboo, etc. Northern tropical dry deciduous forests are an edaphic climax met in areas of varying extent in south facing hill sides, flat hill tops, eroded grounds etc; ascending to the height of 1250 m and sometimes upto 1500 m. These consist of (Bakli) *Anogeissus lotifolia*, with *Buchanania lanzan*, *Sterculia*, *Bauhinia*, *Terminalia tomentosa* with shrubs like *Woodfordia fruticosa*, *Indigofera pulchella* and *Carrisa oppaca*. It can be seen from the southern boundary in the plains to the Tarai and Bhabar, the Dun areas and Siwalik slopes. These forests also penetrate the various inner valleys along their banks about 1400 m. height. These forests are very

Table 14.7: Forest Types in U.P Himalayas

Sl.No.	Forest Types	Altitude	Area (sq. km.)		Dominant species
			Garwal	Kumaun	
1.	Tropical Moist Deciduous forests	300-1500	442.0	239.2	*Shorea robustra* and others
2.	Tropical dry deciduous	Upto 1400	1046.5	12.8	–do–
3.	Sub-tropical Pine forests	1000-1500	2165.14 and others	12.8	*Pinus roxburghii*
4.	Himalayan Moist Temperate forests	1000-3400	4662.6	946.0	*Q. leucotirchophora*, *Q. floribunda*, *Q. semicarpifolia* and others
5.	Himalayan Dry Temperate forests	2300-3200	22.5	0.1	*Hippophae*, *Salicifolia*, and others
6.	Sub-alpine forests	2300-3200	112.3	46.4	*Abies*, *Spectabilis* and others
7.	Moist Alpine Scrub	Above 3100	1383.0	3.8	*Rhododendron* species and others

Table 14.8: Forest Area Lost for Various Purposes in the Himalayan States (000 hectares)

State	*River Valley Project*	*Agri. Purposes*	*Construction of Road*	*Estb. of indust.*	*Mis. Purposes*	*Total*
Assam.	19.7	17.6	6.6	2.9	25.8	72.6
Himachal Pradesh	7.7	12.2	1.2	–	5.7	26.8
J & K	0.1	0.8	0.2	–	90.2	90.8
Manipur	–	–	–	–	–	–
Meghalaya	–	–	–	–	–	–
Mizooram	–	–	–	–	–	–
Nagaland	–	–	–	2.0	0.1	2.1
Sikkim	–	–	–	–	–	–
Tripura	7.9	11.2	0.2	–	19.7	39.0
U.P.	102.0	84.0	4.8	23.4	30.2	245.0
Arunachal Pradesh	0.1	26.3	0.4	7.1	0.1	40.5

useful for timber, industrial wood, fuelwood and fodder and are under very heavy biotic pressure, resulting in extermination and giving rise to barren slopes, mainly along the lower parts of the rivers namely Tons, Yamuna, Bhagirathi, Alaknanda, Nayar, Ramganga, Kosi and Sharda. These forests which cover about 1740 sq. km. are in U.P.Himalayas.

2. Sub-tropical pine forests are found on altitudes ranging from 1000 m to 1800 m. It is also seen on ridges of 600 m. high as 2300 m. on south exposed slopes in the region. These forests cover about 2178 sq. km. area mostly in the reserved forest.
3. The Himalayan moist temperate forests are found in the rainfall zone of 1000 to 1500 mm. per annum. The chief conifers are *Abies picea* with *Cedrus* and *Texces baccata.* Among broad leaved species Oak, *Rhododendron arboreum* and *Pieris ovalifolia* are commonest associates. The pine forests are interspersed with low level Oak. In the neighbourhood of villages where Oaks are subjected to uncontrolled lopping and felling, the pine invades Oak. High level broad leaved species are *Acer, Corylus, Aesculus, Prunus, Carpinus, Juglans* and *Betula* forming consociations. *Arunadinaria* and other shrubs are *Indigofera degmodium* and *Viburnun.* All these forests cover 5600 sq. km. area in the upper Alaknanda, Mandakini, Bhillangana, Bhagirathi, Yamuna and Tons Valleys.
4. The Himalayan dry temperate forests occupy the altitutes above 1700 m. It is an open formation. *Conifer* species include *Cedrus deodara* and *Juniperus* with *Abies* and *Pinus wallichina* at higher elevations. Broad leaved species like *Acer, Froxinus* and *Quercus* are also present. Deodar, being one of the most beautiful trees and providing the best timber, has been extended by plantations since centuries in the entire hill region as temple groves and forest plantations. The lofty Deodar trees around Jageshwar temple (Almora district) Hawalbagh, Dhauladevi, Champawat, Lohaghat etc are some of the examples.
5. Silver-fir and spruce forests form pure forests in Chakrata, Tons, Yamuna and Uttarkashi Forest Divisions. Often

spruce and fir form intimate mixture with Kharsu *e.g.* in Bhillangana Valley.

6. Sub-alpine forests are between 2900 m. and 3500 m. with characteristic dense growth of small crooked trees or large shrubs and patches of coniferous wood. Moist alpine shrub are low evergreen dense scrub with Rhododendron species having some birch and other deciduous trees. Mossess and ferns cover the ground along with alpine herbs. Dry alpine scrubs adjoin dry temperate forest and is xerophytic with dwarf scrubs of *Juniferus wallichiana*, *Coragana* species, *Artemisia* species and herbs.

Table 14.9: The U.P. Himalayas: Forest Areas (sq. km.)

Region	*Poor*	*Medium*	*Good*	*Total*
Garhwal				
Area	2942	3285	1245	7472
% of forest	39.4	43.9	16.7	100.00
% of total area	9.8	11.0	4.1	24.9
Kumaun				
Area	1767	4381	1017	7165
% of forest	24.7	61.1	14.2	100.00
% of total area	8.4	20.9	4.8	34.1
Total				
Area	4709	7666	2263	14638
% of forest	32.2	52.4	15.4	100.00
% of total area	9.3	15.4	4.4	28.7

Degradation

The forests of Himalayan region in the temperate zone, where the bulk of population lives had, suffered due to uncontrolled fellings, shifting cultivation, lopping, grazing and fires annually. Now only 6 to 8 per cent forests are left. The British occupation to reap economic benefit from forests, further deteriorated the situation in the Himalayas. The growth of railways (1853) was phenomenal. Transport of Sal of submonatane tracts from western U.P to Bengal was in great demand for subsequent years. The deodar exploitation climaxed in Kashmir, Himachal Pradesh and Yamuna Valley in U.P with construction of the N. W. Railway line in Punjab, Frontier and Sindh tracts linking Lahore, Peshawar, Karachi, etc.

Table 14.10: The U.P. Himalayas–Forested Area (sq. km.)

District	*Reserve Forest*	*Civil and Soyam*	*Panch-ayat*	*Private and Municipal*	*Total*	*% to Dist. Area*	*% of Regional Forest*
Almora	1490.4	1821	628.0	0.5	3940.9	73.2	13.57
Pithoragarh	1381.3	1210	714.4	–	3305.7	37.7	11.38
Nainital	3563.6	192	207.7	21.1	3984.4	58.7	13.72
Garhwal	2469.2	1806	376.0	6.1	4657.3	85.6	16.03
Chamoli	3704.9	1043	521.5	–	5269.3	57.7	18.14
Uttarkashi	2694.3	870	–	–	3564.7	88.6	12.27
Tehri	1495.7	516	–	205.0	2216.6	80.6	7.60
Dehradun	1954.9	148	–	–	2102.9	71.8	7.24
Total	18754.3	7606	2447.6	232.6	29041.9	69.2	100.00
Per cent	64.54	26.2	8.4	0.8	100.00		

The coniferous forests of Chir, Blue Pine, Spruce and Fir were not lapped due to the lack of transport network. In 1861 forestry started in India. The extension of communication and industrial revolution resulted in over exploitation and complete disappearance of unreserved forests. The passing of 1868 Indian Forest Act and its revision in 1878 led to the reservation and regulation of Sal and Deodar forests which have, thus escaped denudation. Later, the Chir Pine resin became an important raw material for industries and distillation started in U.P.

Establishment of tea gardens, apple orchards and potato etc. as cash crops and horticulture introduction took their toll of the forests. Efforts to establish mining and smelting of iron ore based on wood charcoal before the opening of Suez Canal in 1869 also destroyed the forests. Cantonment towns in hills resulted in the disappearance of forests in their neighbourhood.

The situation in the princely states was no better due to rudimentary system of administration and being treated as the property of the ruler which were gifted as rewards for services rendered. Wilson's exploitation of Tehri state forests, from the first half of the centry, played havoc with deodar of Bhagirathi Valley through river transport is well known. Further with the conquest of Sialkot in 1849, Chenab Valley and Kashmir Valley, deodar was used for the British defences and for building activity in the Punjab plains. Large areas of Pangi in upper Chenab were given on lease to an Amercian for harvesting of deodar by the Raja of Chamba.

A number of forest – based industries are dependent on the Himalayan forests, *e.g.*, resin, rosin and turpentine and their important derivatives like camphor, menthol, rubber emulsifier, paints, etc., match industry, paper and rayon industry, plywood industry, shuttle blocks and bobbins, packing cases and other products, pencils, sports goods, etc. besides timber, railway sleepers, defence purposes, electric transmission poles and firewood. The network of roads in the Himalayan region is approximately 44,000 km. long; many of these roads have been improperly planned and unscientifically constructed leading to destabilization of hillsides and production of colossal amounts of debris. These roads must be producing about 24 x 10^3 m^2 of debris every year in addition to the removal of debris 2650 million tones at the rate of approximately 60,000 m^3/km for road when constructed.

Cattle pressure in the Himalayan region far exceeds the supporting capacity of forests and pastures. There is only 0.32 to 0.060 ha. of grazing land available in Kumaun while it requires 6 ha. per cattle. Due to the grazing pressure, the young regeneration is destroyed, the forests are defoliated by lopping for leaf fodder, the grasslands trampled and their nutritious vegetation smothered, and hillslopes furrowed. The plant cover of the land in the heavily grazed grasslands has dwindled to mere 16.8 per cent at the altitude of 763 m. 24 per cent at 1080 m. 25.85 at 1650 m. in Himachal Pradesh and to 38 per cent at the elevation of 1800 m. in Jammu and Kashmir.

Encroachment of forest land for agriculture has been another factor of deforestation in the Himalayas. Due to the slope productivity of land is low, marginal and sub-marginal lands have also been brought under cultivation leading to more than 30° slope, where sometime ago the forests were developed. Shifting cultivation in northeastern India, is another most important factor produces heavy loss to the forest resources particularly in Arunachal Pradesh and Sikkim. With shifting cultivation soil fertility diminishes rapidly, and with the protective vegetal cover loss, there is massive soil erosion. Shifting cultivators move from place to place destroying forests and as a consequence ecological regression takes place.

Construction of dams over hill rivers to meet the increasing demand of irrigation water and power, of plains, has been taken up and number of projects have been proposed for future. These projects have created a problem of flooding of flat valleys, rehabilitations, loosing of formation, etc. The Tehri dam over Bhagirathi in U.P and Pong dam in Himachal Pradesh have created a threat to the Himalayan environment.

Apart from these, monoculture for industrial plants by clear felling natural ones and some times certain parasites and pathogens may result in conspicuous change in physiognomy of the forest complex. In recent years great interest has been aroused in the medicinal value of many Himalayan herbs and the crude drugs fetch high prices, particularly in the foreign market. This has resulted in their removal from habitats in large quantities.

Deforestation in the Himalayas for various purposes have damaged and destroyed the vegetal cover. The forest cover is left only 18-19 per cent in Himachal Pradesh, 28.7 per cent in U.P. Himalaya and 23 per cent in Nepal. On an average, the forest canopy over the Himalayas has dwindled to mere 35 per cent against the optimum of 60 per cent as per the national forest policy.

The vegetal cover has been stripped of which protects the soil. Now the ground is no longer capable of absorbing rain water, so that much of the precipitation runoff, eroding the hill slopes at furious pace and flooding the down streams. Due to the deforestation capability of the soil to absorb and hold the rainwater diminished, there is no recharge of the ground water and hence, no springs downslope. The rivers will carry less water (20-30 per cent) and only snow-fed rivers would be available.

The U.P. was an independent territory ruled in small kingdoms till as late as the end of eighteenth century. The region was subjugated by Gorkhas in 1794 and by the British in 1815. The smallness of kingdoms, unending conflicts between the local Rajahs, availability of forest area in plenty and thin population, resulted in forests being left to their own. No conservation or management system was prevailing. The forests, were therefore, used in reckless manner to meet daily needs of the local people. People treated forests as their own property for their daily requirements. They were also used as open grazing grounds of large flocks of cattle, sheep and goats. The forests were no man's property and the 'State' also did not care for them. Hence, it became a tradition of annual uncontrolled burning of the forest for getting new grass. Tree felling was unscientific, leaving most of the wood to rot in the forests, repeated uncontrolled lopping for fodder, clearance of forests for herding cattle and extension of cultivation in land even unsuitable for it.

Table 14.11: The U.P. Himalayas–Forest Area Lost for Various Purposes (sq. km.)

Duration	*River Valley Project*	*Agri. Use*	*Road Const.*	*Industries*	*Others*	*Total*
1951-73	932.00	797.00	34.00	193.00	199.00	2155.00
1973-74	0.41	1.78	0.78	0.62	0.13	3.72
1974–75	0.02	38.95	3.33	0.41	1.86	44.57
1975–76	3.12	0.01	7.86	–	3.75	14.74
1976–77	87.32	2.89	1.45	0.23	8.54	100.43
1977–78	–	–	0.69	–	84.03	84.72
1978–79	–	0.07	–	39.99	4.29	44.34
Total	1022.87	840.70	48.11	234.25	301.60	2447.53
Per cent	41.8	44.4	1.96	9.56	12.75	100.00

Prior to the Britishers forests served mainly for extension of agriculture and as refuge in times of war and unsettled conditions. After the Britishers, the impact of communication, transport and industrial revolution resulted in over exploitation. The first British travellers of western Himalaya reported extensive deep lush coniferous forests in the mountains, protected by marshes, mosquitoes and tigers of the dense forests of Tarai. Bishop Reginald Heber's observation in 1824-25 about the treatment of the forests in Bhabar and Siwalik tracts, presages the fate of the Himalayan environment in the subsequent years. He said; " Great devastations are generally made in these woods, partly by the increase in population, building and agriculture, partly by the wasteful habits of the travellers, who cut down multitudes of young trees to make temporary huts and for fuel, while the cattle and goats which browse on the mountains prevent a great part of the seedlings from rising. Unless some precautions are taken, the inhabited parts of Kumaun will soon be wretchedly bare of wood and the country already too arid will not only lose its beauty but also its fertility". These remarks are applicable to the whole region.

Table 14.12: The U.P. Himalayas–Forested Land Use (sq. km.)

District	*Poor Forest*	*Medium Forest*	*Good Forest*	*Total Forest*	*Total Area*	*% of the District*	*% of the Total Forest*
Almora	404	1138	238	1780	5361	33.2	12.2
Nainital	872	1900	254	3026	6780	44.6	20.7
Pithoragarh	491	1373	525	2359	8843	26.7	16.1
Garhwal	826	972	150	1948	5440	35.8	13.3
Uttarkashi	293	491	440	1224	8006	15.3	8.4
Tehri	900	358	330	1588	4423	35.9	10.8
Chamoli	870	1260	69	1999	9003	22.1	13.7
Dehradun	253	204	257	714	3096	23.1	4.9

Source: Singh, 1985.

After independence, the tract, of Tarai, Bhaber, Dun and outer foothills which bears economically valuable sub-tropical deciduous forests have been destroyed. The original inhabitants of the border area have been deprived from winter grazing of their sheep and goats and have stopped their winter migration partly due to clearance of bulk of the Tarai, forests and establishment of huge farms (Kham) by settlers, mostly from west Punjab and partly by the reservation of forests in the foothills and Bhabar region.

The bulk of population in hills live between 900 m. and 1800 m. The density of population is high with more area under cultivation and hence, leaving less area under forests. The Government policy has failed to fulfil the assurance given to the local people regarding the forest conservancy and at the time of reservation of forest. It was natural for the people to resent their forests being used for supply of industries elsewhere without meeting their basic needs. The lifesytle and misery of their life continued unchanged and unaffected by the development of the modern technical age. Masters and managers, manage by remote control and are unaware of the hard life of hill-man. Still the women folk have to arrange for daily needs of family and large herds of cattle. They are even now living in the same lifestyle as their forefathers used to do centuries ago.

The impact of Five Year Development Plans is noticeable in the form of motor roads, bridges, schools, buildings, pipe lines, electric lines, communication lines, etc. and large number of educated young men in search of white collar jobs, unemployed or migrated leaving female children and old people to look after their belongings. The agriculture is still transitional and unable to feed them. The carrying capacity of land is only to feed 55 per cent of the population and 35 per cent of cattle. The density of population is 91 persons per sq. km. and land-man ratio as calculated to be only 0.15 ha.

The U.P. Himalaya has 65.45 per cent area under forests of the total geographical area of the region, 45.13 per cent area is under the Forest Department, 20.32 per cent is under community forest and private control. Out of the total forest area 64.45 per cent is under reserved forest, 26.2 per cent Civil Soyam forest, 8.4 per cent Panchayat forest and under Private, Cantonment and Municipal forest is 0.8 per cent. The community forests extending to 10,000 sq. km. have completely disappeared. Class I forests have also deteriorated being with revenue authorities.

The Sal zone of the outer hills, Bhabar and Tarai is now threatened by the silent process of destruction due to lopping and grazing by Gujjar buffaloes. In winters large flocks of sheep entering these forests, coming even from the farthest regions of Himachal Pradesh, decimate whatever natural regeneration that happen to come and escape the summer fires. In the process of development Himalayas have lost about 5 per cent area (2447.53 sq. km.).

The Chipko Movement took place in 1974 in the region; which had originated in Rajasthan sometime in the eighteenth century about 200 years ago by Bishnois, who resisted felling of trees for the brick kiln of a local king for building his palace.

The present state of forests in the U.P. can best be understood from the Table 14.11. It shows that the region has good forest in 4.4 per cent (2263 sq. km.) of the total geographical area (50952 sq. km.). The crown cover is between 30 and 60 per cent in 7666 sq. km. area while it is 10 per cent–30 per cent in 4709 sq. km. The Garhwal

Himalaya in region has less area under forest (24.9 per cent) and Kumaun Himalayas has more (34.1 per cent) to the total geographical area. As much as 36.8 per cent of the geographical area has been declared under forest but the crown cover is less than 10 per cent.

Forested area in the Uttarakhand Himalaya vary from 4.46 per cent in Nainital to 15.3 per cent in Uttarkashi from the total area of the district. Out of the total forest area 20.7 per cent is found in Nainital while it is only 4.9 per cent in Dehradun.

References

Additional Director Agriculture (Statistics) U.P., 1989. Data for Economic Regions, Directorate of Agriculture, U.P. Lucknow.

Anonymous, 1978. Forest area lost for various purposes in India; Statewise 1951–52 to 1975–76. *Voluntary Action*, 20(12): 29.

Champion, H.G. and Seth, S.K., 1986. *A Revised Survey of Forest Types of India*. Manager of Pub. Govt. of India, Delhi, P.404.

Government of India, 1976. Report of National Commission on Agriculture, Part IX Forestry, Ministry of Agri., New Delhi.

Gupta, R.K., 1983. *The Living Himalayas*. Today and Tomorrow, New Delhi, p. 56.

Mannering, J.V., 1981. The use of soil tolerance as strategy for soil conservation: Problems and prospects. (Ed.) R.P.C. Morgan. John Wiley and Sons, Chichester, England, p. 337–349.

Narayana, V.V.D. and Babu Ram, 1983. Estimation of soil erosion in India. *Journal of Irrigation and Drainage Engineering*, 109(4): 419–434.

Omaston, A.E., 1921. Working plan for the North Garhwal Forest Division 1921–22–1930–31, Allahabad, p. 35.

Seth, S.K., 1978. Forest and forestry in the Himalayan Region. In: *Proc. Nat. Seminar, Resource Development and Environment of the Himalayan Region*, New Delhi, pp. 70–97.

Singh, J.S. and Saxena, A.K., 1980. The grass cover in the Himalayan region. In: *Proc. Nat. Seminar on Resources*.

Stebbing, E.P. *The Forest of India*, Vols. I to IV.

Stowell, V.A., 1907. *A Manual of the Land Tenures of Kumaon Division*. Govt. Press, Allahabad, pp. 143–144.

Teli, B.L., 1983. Levels of agricultural efficiency in India. In: *OISCA International Symp.* by International Union of Cons. of Nature and Natural Resources, Tokyo, Japan.

Teli, B.L., 1987. Himalayan land use structure: Change planning a case study. In: *Proc. of Int. Symp. on Changing Agri. Rural Dev. in Developing Countries*, B.H.U., Varanasi.

Teli, B.L., 1988. *Himalayan Land Use Structure: Planning and Potentialities*. NATMO Monograph No. 81, Calcutta, pp. 56–60.

Teli, B.L., 1992. Environment people and land in the Himalayas: Problems and planning. In: *The Himalayas Heritage*, (Ed.) K.C. Purohit (In press).

Teli, B.L., 1992. Patterns of variations in the agricultural development of the Garhwal Himalayas. In: *Himalayan Environment: Man and the Economic Activities*, (Ed.) J.L. Raina. Pointer Pub., Jaipur, 2: 369–376.

Teli, B.L., 1992. Ecology, farming systems and cropping pattern. In: *Himalayan Environment: Man and the Economic Activities*, (Ed.) J.L. Raina. Pointer Pub., Jaipur, pp. 147–163.

Teli, B.L., 1992. Rural land scap management and planning in W. Nayar Basin. *Unpublished D. Phil Thesis*, H.N.B. Garhwal Uni., Srinagar, Garhwal.

Teli, B.L., 1992. Management and planning of forest resources in the U.P. Himalayas. In: In: *Himalayan Environment: Man and the Economic Activities*, (Ed.) J.L. Raina. Pointer Pub., Jaipur, pp. 251–269.

Tiwari, A.K., *et al.*, 1985. Inventory of forest biomass for Indian Central Himalaya. In: *Environmental Regeneration in Himalayas*, (Ed.) J.S. Singh, Nainital, pp. 236–247

The Forest Fact-Finding Committee Report, 1959. Government Press, Lucknow.

Valdiya, K.S., 1983. Accelerated erosion and landslide prone zones in the Central Himalaya region. In: *Environmental Regeneration in Himalaya*, (Ed.) J.S. Singh. Ganodaya Prakashan, Nainital, pp. 12–38.

Chapter 15

Study of the Effect of Mining on the Jainti Basin in Eastern Himalaya*

S.R. Basu, Nandini Dhara, Nabanita Sen Gupta and Madhumita Sinha

ABSTRACT

The upper Jainti Basin has rich dolomite deposits which have been extensively exploited since 1931, particularly to cater for the Iron and Steel industries of eastern India. Although commercially viable and necessary for our growing industrial scenario yet, the unscientific mining operation has led to an adverse effect on environment which will lead to loss of biodiversity if not checked in time.

Introduction

Nature is bountiful. It generously bestows its immense wealth to man, if he lives in concert with nature. But when he does not, the products of his labour and sometimes his very existence are threatened to be obliterated. The area under study presents a glaring example of such a possibility where unscientific mining and deforestation with the sole aim of maximization of profit has defaced and spoiled the environment to disastrous and alarming proportions.

The mining area comprises of the region immediately north and north-east of Jainti village, at the south-western, southern and the eastern slopes of Mohakal hills between the parallels of latitude 26° 40´ N to 26° 44´ N and longitudes 89° 41´ E in the Survey of India

---

* *Chapter Courtesy*: Vinod Publisher & Distributors, Jammu.

Toposheet No. 78 F/10. The border of West Bengal and Bhutan runs along the crest of the Sachaphu Ridge and climbs down westward to the Jainti river (from which the village derives its name) in a north and north-westerly direction.

Physiography of the Area

The area forms a part of the sub-Himalayan zone which gradually gains elevation from the southern alluvial plain to the hilly terrain in the north. The general slope varies from 30° to 40° in the lower portion to nearly 50° in the Upper part and to almost precipituous scarps towards the summit of the ridge. Two broad divisions may be made–(*i*) high hills comprising of Buxa and Gondwana rocks in the north and (*ii*) low hills comprising of Siwalik rocks to the south.

The drainage in this area is in general southerly. A number of 'nallah' (small tributaries) with steep gradients drain the hilly terrain and join river Jainti to the south. Most of the rivers formed by these 'nallah' are non-perennial in nature except the river Jainti.

Geological Background

A generalized geological account of the area is as follows:

Pleistocene to Sub-Recent and Recent

Deposits of unstratified boulders and gravel drifts occupy the lower slopes and the plains at the foot of the hills. The area further south is covered by alluvial deposits.

Tertiary (Siwalik)

Fine grained friable and soft micaceous sandstone intercalated with pebbly sandstone and shales with lignite and coal belonging to Siwalik system (Nahans) of Tertiary age are to be encountered further to the north of Pleistocene deposits.

Gondwanas (Permo-Carboniferous)

Fine to medium grained calcareous sandstones and quartzitic sandstone interbedded with thin slaty beds come next to the Siwalik system in the north.

Early Paleozoic

A fairly wide belt of crystalline grey dolomite beds, quartzite schist and phyllite, forming the Buxa series (Purana), occupy the southern slopes and entire north-western part of the Sachaphu ridge and Mohakal hill.

Archaeans

Daling series consisting of slates, phyllites, banded haematite quartzites schists and Darjeeling gneiss are to be found in the upper most part of the basin.

There are a number of major thrusts in the area which have caused reversal in the sequence of formations, the dolomite of the Buxa-series forming the higher ridges over the younger Tertiary sandstone. The most important thrust plain is the Main Boundary Fault which separates the Gondwanas in the north from the Siwaliks in the south. Further north, the Gondwanas are separated from the Buxas by the Buxa-Thrust (Kurien, 1962).

The general strike of the rocks is from north-east to south-east and the average dip is from 40° to 50° to the north-east. Four to eight sets of joints are well developed in quartzite and dolomite.

Dolomite Exposures

The region to the north and north-east of the Jainti village is well known for its high quality dolomite and limestone. Limestone containing more than 20 per cent magnesium is known as dolomite. The magnesium content of dolomite in this region varies between 20-21 per cent.

According to D.K. Chandra, Assistant Geologist, Geological Survey of India, 8 sq. km. out of 11 sq. km. of dolomite occupying area falls within the Sachaphu Reserved Forest on the side of the River Jainti in West Bengal and the remaining is in Bhutan. The exposure has a width of about 800 m. On the eastern slopes of Mohakal hill, the outcrops of dolomite being aligned north-west to south-east to the north and north-west of Pashkhoa Tea Garden. The total reserve in the area runs into many million tons. D.K. Chandra collected 16 samples of dolomite and their chemical analysis reveal that SiO_2, Al_2O_3 and Fe_2O_3 is less than 1.0 per cent in each and MgO is above 21.0 per cent in the majority of cases. The entire width of the belt appears to be composed of almost pure dolomite which is suitable for flux and lining in Iron and Steel furnance, in the manufacture of metallic magnesium, epsom salt, high magnesia cements etc.

History of Mining

Dolomite mining in the Jainti basin like the other mining areas in India has grown up in a backward area in the dense Sachaphu Reserve Forest. Mining first started here in 1931 with establishment of Bengal Lime and Stone Company. Formerly there were four mining and quarrying companies, *viz.*

1. Bengal Lime and Stone Company Private Limited
2. Buxa Dolomite Company Limited
3. North Bengal Dolomite Company Limited and
4. Jainti Lime Company

At present out of the above four companies only two are successfully working. They are:

1. Bengal Lime and Stone Company Private Limited and
2. North Bengal Dolomite Company Limited

Buxa Dolomite Company and Jainti Lime Company are no more working due to their uneconomical conditions.

Legal Aspects of Mining Dolomite in Jainti

Land is taken up on lease for mining dolomite. Dispute prevails over the grant of mining lease in the Sachaphu Reserve Forest. Bengal Lime and Stone company holds a quarry lease for limestone near Jainti under the Bengal Waste Land Manual Rules. In 1941 the company took a prospecting licence for dolomite over an area of 21sq.kms in this region and in 1945 they applied for a mining lease for dolomite for the manufacture of metallic magnesium over an area of 8sq. kms within the limists of their prospecting licence.

Objections were raised by the Divisional Forest Officer, Buxa-Division to grant this lease, as mining would have invariably led to erosion and landslides leading to possible danger to the bridge on the river Jainti the Jainti-Dhowla road the railway sidings, stone quarries, forest and tea gardens further down. The tract composed of soft and friable sandstones are easily eroded during floods and are likely to cause landslides. So it was not in the interest of the forest and the safety of the neighbourhood to lease out this area (Narayanswami, 1951).

One on the eastern slopes of Mohakal hill carries fairly large reserves and mining of the dolomite from this area would not have affected the stability of the hill sides in view of the hard nature of the rock and absence of loose textured Tertiary sandstone and so there was no objection on issuing mining lease over this area.

A fairly wide extensive belt of dolomite occurs on slopes immediately north and north-west of Pashkhoa Tea Garden. The southernly projection of the peak 3769′ between the upper reaches of Chuniakhola and Pashkhoa nallah was admirably suited for mining of dolomite on a large scale without any fear of causing damage and as such the Government had no objection on grounds of safety of hill sides and forests for issuing lease.

Bengal Lime and Stone company started operations by picking up and blasting dolomite boulders from the stream courses of Majakhola, Dumilekhola and Bajekhola. But now their exploitation of this resource is much more extensive and detrimental to the environment.

Operational Processes in a Dolomite Quarry

On visiting a quarry site one is struck by the crude and unscientific way in which quarrying is done and by the sheer negligence to protective and restorative measures on part of the authority concerned.

The sole motive of profit maximization has led to the use of pulverizers, crushers, generators and trucks but there is not even a semblance of effort for the safety of the workers or to undertake quarrying in a scientific way.

Quarrying is a seasonal operation usually from October to April. A suitable area easily accessible and having rich and good quality dolomite deposits is chosen, often by experts from Tata who have their vested interests. A narrow unpaved pathway is made on the slopes to reach the deposits and a crude framework of bamboos is made for the purpose of transferring dolomite from the hill slopes on to the ground. Workers often fall to their death from these high unstable structures and their families are hardly given any compensation. Labourers without having any head gear reach the dolomite deposits and dislodge them with shovels allowing them to fall freely to the ground creating lot of dust leading to environmental pollution.

The crushers and pulverizers break the dolomite chunks to various sizes and are also sieved to be graded into dust grit chips and lumps to be used for different purposes. So there is almost hundred per cent utilization of dolomite with minimum effort and capital investment. Profits are high as there is minimum expense on labourers and hardly any on restorative measures.

The workers live in makeshift dwellings near the mining sites without any proper amenities. Sometimes they fall prey to the tigers of the jungles and hardly any compensation is paid to the families. They are hardly provided with protective headgear such as helmets when they are working. A striking fact is that most of the workers are not of local origin but brought in batches from Bihar and Orissa by agents seasonally.

Once the deposits are fully exploited at a spot they move on to a new spot. The abandoned quarries are neither stacked nor replanting done on the deforested and scarred slopes. The mine shafts and gaping pits collect water encourage gullying and ultimately lead to the formation of badlands. Breached portions of the roads are poorly treated by dumping dolomite. Thus there is no vestige of scientific approach to the operational processes in the dolomite quarries, the only motive being the maximization of the profit at the expense of nature and under priviledged poor workers.

Case Studies

Bengal Lime and Stone Company Private Ltd.

Bengal Lime and Stone Company started on 1931. It was founded by a renowned geologist Mr. M. K. Roy who first discovered vast reserves of dolomite in the Jainti basin.(Kusari 1981)

Bengal Lime and Stone Company operates two quarries. One is located on the eastern side of the Sachaphu Reserved Forest where the lime stone quarries are situated. The other is located on the north of the Jainti Dolomite mine. The total leased area occupies about 310 acres. There are five pits from where dolomite is mined. The mining season starts from October and ends in the month of April every year. A large labour force is recruited from the neighbouring states of Bihar and Orissa as seasonal labourers. The mining labour in Bengal Lime and Stone Company is about 150, out of which 19 are permanent at Jainti factory. 4 departmental trucks carry dolomite from mines to stackyard.

There are 4 labourers working in each truck and the truck makes about 8 trips a day. Their pay increases with the size of the dolomite (Table 15.3). One truck carries about 8 tons of dolomite per trip. It is sold at approximate rate of Rs 100 per ton. When compared to the amount sold the labourers get only a meager amount. They are not paid on a monthly basis but on the amount of work done per day.

21 office staff are working in the office at Jainti. The head office is at Kolkata. Yearly production of dolomite in Bengal Lime and Stone Company Private Limited ranges between 6,517to 30,000 metric tones (Table 15.1).

Table 15.1: Production Statistics of Bengal Lime and Stone Company Private Limited

Year	*Production in Metric Ton*
1982–83	10,647
1983–84	6,517
1984–85	29,846
1985–86	30,000
1986–87	24,471

There are five types of grading (size) in dolomite. They are: (*i*) Dust which is chiefly used as fertilizers in Tea gardens and for agriculture as it is rich in magnesium, (*ii*) Grit which is exported to Chittagong Steel Plant, (*iii*) Chips which are put into use by Durgapur Steel Plant, (*iv*) Tata which is sent to Tata Iron and Steel Company and (*v*) Lump generally used for the production of lime (Table 15.2).

In 1980 the company had set up a machine for increasing the dolomite production. Daily production with the help of the machine was about 50 metric tones. But now due to the objection made by the Forest Department that any type of mining in the Tiger Reserve Forest destroys the environmental balance and as a result in 1986-87 the production of dolomite dwindled to only 24,471 metric tones (Table 15.1).

Table15.2: Grades of Dolomite and Use

Sl.No.	Grade	Size	Use
1.	Dust	0.023 mm	Chiefly used as fertilizer in Tea Gardens and Agriculture
2.	Grit	0.327 mm	Exported to Chittagong Steel Plant, Bangladesh.
3.	Chips	0.725 mm	Goes to Durgapur Steel
4.	Tata	0.725-0.750 mm	Sent to Tata Iron and Steel Plant
5.	Lump	4"-6"	Used for the lime production

Table 15.3: Payment of Labourers

"Grit" per Lorry	Rs. 16/-
"Chips" per Lorry	Rs. 30/-
"Tata" per Lorry	Rs. 60/-

The problems faced by this company are many. The first and the foremost is the road maintenance as there is no proper road system connecting the mining sites and nearby towns. Therefore the company has to maintain the road. The road is subjected to problems like landslides and erosion every year.

The second problem is the supply of electricity. Since there is no generating unit nearby the company has to generate its own electricity with the help of generators.

The third problem which the company has to face is the lack of transport. Being situated in remote places the only mode of transportation of its product is through trucks. Formerly there was a meter gauge railway line connecting it with other places to transport the dolomite from the mine pits but at present that too has stopped. Now the company has to depend solely on trucks.

Royalty paid to the central Government is about Rs 5.00 per metric ton of dolomite.

North Bengal Dolomite Limited

North Bengal Dolomite Limited started in 1978 is situated about 15km. east of Jainti Hatipotha mines. It is the only public sector company in that area, under a joint venture of Steel Authority of

India (India Government Undertaking) and West Bengal Mineral Development and Trading Corporation Limited (West Bengal Government Undertaking). The total mining leased area is about 8.97 sq.km.

The mining season continues from November to March. Labour force is mainly recruited from Orissa. The total official employees of the company are 45. The dolomite is stored at Phaskhoa. From there it is carried to Alipurduar by trucks on a contract basis. Monthly production of dolomite in North Bengal Dolomite Limited (Hatipotha mines) varies between 1,831.92 to 13,772.55 metric tonnes. The total production of dolomite in 1986-1987 was about 40,000 tonnes. There is no blasting process in this mine. Dolomite produced in this mine is extensively used in Durgapur Steel Plant, Bokaro Steel Plant and in agriculture fields as fertilizer.

Effect of Mining on the Environment

The unscientific and haphazard mining of dolomite in the hill slopes in the upper Jainti basin has led to disastrous effects related with (*a*) Soil-erosion, (*b*) Land-slides and (*c*) Floods.

Soil-erosion

Deforestation as a result of clearings for mining and quarrying operations lead to large scale soil-erosion ultimately converting large patches of forest land to useless badlands (Basu and Ghatowar, 1988). After extraction of dolomite, the slopes are never treated properly, neither is replanting done to arrest soil-erosion.

Landslides

Rain water collects and forms under-ground channels through abandoned mine shafts which are never stacked or filled with sand after total extraction and leads to disastrous landslides and slumps. Soil slides are generally of small magnitude mainly occurring along the lower slopes while debris slides are generally of greater magnitude and are quite common and devastating. When saturated debris along with big blocks of rocks, flows downward to form a channel of its own, mud-rock-flow takes place. A large part of the area is composed of soft and friable sandstone which are easily dislodged during mining operations and later cause large land-slumps. The streams draining the area cut deep gorges into the sandstone and the valleys have extremely precipitous walls. The valleys of Mathekhola and Kalikhola are highly prone to landslides.

Floods

Excessive soil-erosion and landslides lead to an increase in the load of the rivers. The heavy input of the debris is deposited at the foothills as there is a sudden change in gradient. The added material also raises the bed of the river and many bars and shoals are formed. Coupled with these are the depleted forest cover and the heavy and concentrated rainfall varying between 350 cm in the south to 500 cm in the north which leads to devastating floods. As the washed in materials from dolomite dumps, soil-erosion and landslides have already choked the streams, a little extra amount of discharge as in the monsoon months leads to disastrous floods resulting in loss of life and property and this has become a regular feature. Severe bank-erosion and uprooting of trees is also the result of bankful conditions and floods. The North Bengal Flood Control Commission formed to control floods in this area has not yet done any substantial work in this part of the Jainti Basin and so the problem of floods is ever increasing.

Impact of Mining of Dolomite in the Natural System

To make an environmental impact appraisal, the natural elements of the study area have been carefully identified. The various cause and effect relationships between different human activities and the natural systems and the resulting environmental impact have been summarized as follows in Table 15.4.

Conclusions and Recommendations

Dolomite is being actively quarried at present by the Bengal Lime and Stone Company Limited and the North Bengal Dolomite Limited.

The present study reveals that haphazard and unscientific mining operations destabilize slopes, leads to excessive soil-erosion, gullying, formation of badlands and landslides, generates dust and noise thereby polluting the environment. Materials from slumps, landslides and dolomite wastes choke up the streams leading to floods. All these spell disaster for the region. Steps of amelioration of adverse effects may be taken by excavating short trenches parallel to contours with subsequent grading, construction of contour parallel

Table 15.4

A. Deteriorative Effects Due to Mining

Sl.No.	*Natural Elements*	*Human Interference*	*Effect*	*Ultimately impact on Environment*
1.	Water	(a) Washed in from dolomite dumps	(a) Poor in quality for agriculture and drinking	Deterioration
2.	Streams	(a) Deforestation (b) Siltation by washes from dolomite (c) Abandoned mines left without being filled up	(a) Bar formation (b) Raising of the river bed and consequent floods (c) Floods	Deterioration
3.	Soil	(a) Digging (b) Quarring (c) Mining (d) Open excavated pits	(a) Gullying (b) Slope failure (c) Landslides (d) Badlands	Deterioration
4.	Air quality	(a) Extraction of dolomite (b) Vehicle movement (c) Dumping (d) Seiving (e) Gradation	(a) Dust generation	Air Pollution

Contd...

Table 15.4–Contd...

Sl.No.	Natural Elements	Human Interference	Effect	Ultimately impact on Environment
5.	Noise level	(a) Heavy machineries (b) Loading (c) Transportation (d) Operation of crushers and pulverizers (e) Dumping	(a) Noise generation	Harmful to health
6.	Agriculture	(a) Encroachment (b) Washed in dolomite deposited on cultivated lands	(a) Poor yield and quality	Deterioration
7.	Wildlife	(a) Quarry operations (b) Clearing forest areas (c) Noisy machineries (d) Killing (e) Road construction	(a) Disturbance	Depletion of Biodiversity
8.	Forest and ecosystem	(a) Deforestation (b) Hazardous mining (c) Landslides (d) Badlands (e) Floods	(a) Deforestation (b) Soil erosion	Degradation and Imbalance

Contd...

Table 15.4–Contd...

B. Beneficial Effects due to Mining			
Sl.No.	*Attributes*	*Causes*	*Ultimate Impact*
1	Trade and Commerce	(a) Commercial use of dolomite for iron and steel industry (b) Modern constructions (c) Fertilizers	Augmentation
2	Employment	(a) Quarrying and mining operations (b) Transportation (c) Tertiary sector	Augmentation
3	Tourism	(a) Scenic beauty (b) Dense forest cover with wildlife (c) Pleasant climate (d) Buxa-Tiger project	Augmentation

roads, taking measures to protect slopes by constructing retaining walls in critical stretches, diverting drainage lines, constructing pollution control channels, stacking of mine shafts by sand, replanting plants on the deforested slopes etc. Unfortunately none of these have been implemented till now.

The mining companies specially the private ones in a bid to maximize profit with minimum effort and expenditure have damaged the environment. But their ambitious plans have not benefited the local people as they are not even employed in the quarries. On the other hand they face the furies of nature in the form of landslides and floods. There is no such large unit as a Cement Factory which would have led to industrial development of the region. The life of the local people worsens day by day. The government also does not earn a substantial revenue in the form of royalties. So it is seen that the destructions far outweigh the benefits. Till now the restorative measures have never been taken up and there is much doubt about their being implemented in the near future. So for the ecological balance and the scenic beauty of the region it is better for these hazardous mining operations to be stopped.

To check sound pollution and destruction of virgin forests the railway route through the forests had already been abandoned. The Buxa-Tiger Project has been undertaken by the government to discourage deforestation and mining in the reserved forests and it is hoped that it will be eventually successful in retaining the ecological balance of the region.

References

Basu, S.R. and Ghatower, L., 1988. Mining and environment: A case study in the Lish basin of Eastern Himalayas. In: *Mining and Enviornment in India* (Ed.). HRG Publication, pp. 318–325.

Chandra, D.K. *Unpublished Report on the Dolomite and other minerals in Buxa-Duars, District Jalpaiguri, Bengal and partly in Bhutan territory*, G.S.I.

Kusari, A.B., 1981. *West Bengal District Gazetters, Jalpaiguri District.*

Kurien, T.K., 1996. Geological mapping in parts of Jalpaiguri and Darjeeling Districts, West Bengal, *G.S.I.* Report for 1961-62.

Narayanswami, B., 1951. Unpublished report on the advice for granting mining lease for dolomite to Mr. M.K. Roy in the Sachaphu Reserve Forest, Jalpaiguri District, West Bengal.

Chapter 16

Landslides in Darjeeling Himalayas

Partha Pratim Roy and Jyotirmoy Sen

ABSTRACT

Among the various physical processes, which disturb the ecosystem of the Darjeeling Himalaya, landslide is of vital importance. It not only hampers the ecosystem, but also the socio-economic development of the area. Depending on nature of slope, geology, soil cover, chemical composition, vegetal cover, hydrology, drainage system, settlement and development of road network, frequency and time of landslide will vary.

In the last two decades, Darjeeling Himalaya experienced near about 42 slope failures. Heavy and continuous rainfall, deforestation, urbanisation, increasing road density triggers the mechanism of landslide. Now different techniques have been used to lessen the intensity and frequency of landslide hazard.

Introduction

The Mountain Environment

Mountains are the outcome of the geological and geomorphological processes. Nearly one-fifth of the world's land surface is under mountainous topography. Weak geological structure, steep and rugged surface, high altitude variation and seasonal monsoon rainfall with occasional cloudbursts result in the high degree of fragility in the whole Himalayan Territory.

In modern day, different techniques are applied in geographical research, which not only delineated the affected areas, but also give us the measures to check the hazard. So, to study the landslide hazard in the mountainous. areas, it is necessary to delimit the prone areas, measures the factors and the correlation exists between them.

The Darjeeling Himalayas

The district of Darjeeling lies between 26°31′ and 27°13′ north latitude and between 87°59′ and 88°53′ east longitude. It contains a total area at 1164 sq. miles and a population of 1,605,900 souls.

It is geologically a complicated system and traversed by numerous hilly stream and rivers. It is being compared neither of the alluvium of the plains nor of the rocks of the hills but for the most parts of alternative beds of sands, gravels and boulders brought down by the rivers. In 1814, Mallet distinguished two main lithological units which he termed the Daling series and Darjeeling series.

Darjeeling Series

This is mainly composed of foliated gneisses, mica schists and occasional bands of flaggy quartzites. It proceeds along some susceptible bands, *i.e.*, mica-rich bands in preference to quartzose bands and also along the shear planes. As a result, blocks of fresh rocks are generally found encircled by highly weathered rocks of the nature of clay. The clay is found mixed up with grains of quartz, feldspar and flakes of mica. This has got an important bearing on the slips.

Daling Series

South of Darjeeling Gneiss lies the Daling series. It consists of mainly metamorphosed argillaceous rocks represented by purple slates, quartzites and crumpled phyllite having generally a northern dip. Due to the presence of Daling phyllites, which are very soft and susceptible to erosion, the frequency of landslides is very high.

Physiography

The area of the district is not marked by the natural features as a region complete in itself. It consists of a portion of the outlying hills of the lower Himalayas and a stretch of territory lying along the base of the hills known as the Tarai. The range of altitude is considerable. The Terai is only 300 ft. above sea level but there are parts of the district in the hills which are nearly 12000 ft. high.

From the preceding account it will be seen that Darjeeling falls naturally into two distinct tracts, the Tarai and the Ridges and Deep valleys of the lower Himalaya. The Tarai portion is a low-lying belt traversed by south flowing rivers, formerly covered by dense malarious jungle. Now this belt is extensively used for tea gardens and settled tillage.

Behind the Tarai, the mountains tower abruptly from the plains, which are here only about 300ft above sea level, in lofty spurs reaching to 6000 and 10,000 feet, and culminate in a series of long ridges and valleys.

The Study Area

The study area is very small compare to the total area of the district. It lies in the north-central part of the district, and bound by longitudes 88°20′E to 88°28′E and latitudes 27°0′N to 27°5′N.

The study area has disturbed due to recent slope failure, landslides and tectonic activities. The basin has a catchment area of 26.64 sq.mile. The main stream in this catchment area is the Tista and several other streams or channels join to this main stream throughout the study area.

Geomorphic Relationship with the Adjoining Area

It has a great diversity of landscape with many striking contrasts of landforms and morphogenetic conditions. There is a series of ridge and valleys arranged in an irregular order ascending from north west to south- southeast from middle to eastern and western margins.

This antecedent drainage relating to the Himalayan upheavel with gorge and deeply dissected topography has been carved out of gneisses, and granites, schists, slates, shales and other softer sedimentaries. The wide river beds with heavy deposition of silt and detritus etc. in the lower reaches of the Tista, particularly beyond the limit of 300 ft contour are marked by lateral shifts and subsequent diversions of course, raising of river beds, spilling over banks resulting in heavy floods.

Objective of the Study

The main objectives of this study is to analyse the various natural and social phenomena and suggest various mitigating measures. The objectives are as follows;

1. To assess the nature, type and distribution of mass wasting hazard in the study area.
2. To evaluate the relationship of mass movement hazards and controlling variables (geology, geomorphology climate, hydrology, topography, vegetation and soil)
3. To classify the basin areas in different slope categories.
4. To suggest measures to mitigate the problems of mass wasting.

Methodology

1. Using the basic map on 1 : 63,360 scale, for carrying out the layout maps for the dissertation.
2. Both primary and secondary data are taken into consideration for the dissertation work.
3. Morphometric analysis was taken into consideration to analyse the problem and its measures.
4. Various indices are used to illustrate the different categories of hazard and risk map.
5. Diagrams and photographs related to the disseration work have been presented.

Limitation

It is' not fulfill in its requirements. It has some limitations. The toposheet no. 78 a/8 is restricted in character so proper interpretation will not be possible. As it was surveyed during 1925–26, so recent changes could not be visible in that map. Due to the lack of present maps and satellite imagery proper interpretation of the study area is not possible.

General Background of the Study Area

Geolithology

The main stream is the Tista which originates from Tista Kange or Pauhunri glacier near Kangcheng Lake (27°59′N, 38°48′E, 7128 m.a.s.l.). Along its course we can find a complexity in its geolithology.

In the study area *i.e.*, in the middle Tista Basin we get a good expression of the Siwaliks. In the foothills area these are over thrust by the Damudas. This Damudas are highly sheared and tectonically active areas, exhibiting a greater deformed relic with more reduced section. It constitutes a very important part of Gondwana succession.

So, the 3 major rock formations are the Siwaliks, Damuda and Daling which are separated from each other by the next period of weathering and erosion, earth movements or tectonic activities. The south facings breaks (escarpment) of the rugged hilly terrain is conspicuous and delineates the boundary between the upper Tista basin area and the lower Tista basin.

Physiography

Major portion of this area manifest prominent ridge and gully topography associated with discontinuous gullies of accelerated erosion and landscape instability etc.

The terraces, dissected valley side slopes, landslides slopes, hill-side slopes, truncated spurs are the resultant forms of the fluvial erosion. These terraces are formed by the landslide materials and the related slope wash derived from those adjacent hill side and valley side slopes.

Climate

The basin is marked by its linkages with the past climatic changes *i.e.*, cold, warm, humid and semi-arid conditions of the tertiary and quaternary period.

The terraces of this area are mainly alluvial in nature, while the present day point bars are gravelly, which indicates the variation in discharge due to the climatic variation from the past. Present day discharge is greater than the past rates.

Intense corrasional activity is found due to the occurrences of flood erosions and heavy rainfall (3000 mm–4600 mm in the avg.) being accompanied by tectonic in one hand and human activity like construction of dams, buildings, deforestation on the other.

The local climate depends largely on the elevation which rises abruptly from 300 m to 3600m. The monsoon or the rainy season is the most dominating in Darjeeling, characterized by mist and continuous rain which starts from midway and continues upto

September. The monsoon is followed by a short period of autumn, and from December to late February continues the cold and unpleasant winter.

Rainfall

During the cold weather months very little rain falls in the hilly part. November and December are almost rainless but light showers fall in January and February due to shallow depressions passing eastwards over the plains.

In May, the southerly winds from the Bay reach the hills and caused increased precipitation, which is sometimes very heavy, specially in the later half of that month. Very heavy rainfall is general from June to September, especially on the lower slopes. 80 per cent of annual rainfall occurs in this time accompanied by thunderstorm. Hailstorms occur during the months of March, April and May.

Table 16.1: Normal Monthly Distribution of Rainfall (Excluding Snow and Hail) in Inch

Jan.	*Feb.*	*Mar.*	*Apr.*	*May*	*June*	*Jul.*	*Aug.*	*Sep.*	*Oct.*	*Nov.*	*Dec.*
0.53	0.19	1.88	4.14	9.63	24.18	32.02	26.56	18.90	5.41	0.80	0.27

Temperature

Climatic conditions are necessarily varied more especially as regards temperature. Temperature is almost lower in respect to the Darjeeling towns and other areas like Siliguri, Bagdogra etc. This is because of height and as well as the wind blowing over the river. The winter mean temperature is 40.2°F and summer is 64.8°F

Table 16.2: Maximum and Minimum Temp. (°F)

Jan.	*Feb.*	*Mar.*	*Apr.*	*May*	*June*	*Jul.*	*Aug.*	*Sep.*	*Oct.*	*Nov.*	*Dec.*
					Maximum						
46.6	48.4	56.3	62.5	64.0	65.5	66.3	65.9	64.9	61.3	55.1	49.2
					Minimum						
34.7	35.5	42.1	48.5	52.1	56.2	57.4	57.4	57.7	49.9	42.7	36.7

Frequent norwesters, and some local disturbances may sometimes reduce the temperature upto 35°F. But throughout the year, the area is characterized by higher temperature range of 50°F.

Humidity

The atmosphere is highly humid throughout the year, the Relative Humidity ranges between 90–95 per cent during rainy season.

Cloudiness

The sky is heavily overcast and is not discernible for 20 days on the average in each of the months of July, and August, but it is high to moderately clouded in other months. This time mean temperature is steady at or slightly above 60°F.

Winds

In March a brief Himalayan spring is ushered in with high blustering winds; it is very short, lasting only till the end of the month, but during this time the slopes lose the brown dried-up appearance which they get in winter. During monsoon, the wind direction current occasionally rises to great heights. Surface wind have usually an easterly compound, but in winter the wind direction is towards east and north-east.

Soil

Greater portion of the underlying rock consists of slates, schists and quartzites. But in few places, the underlying rock consists of gneiss occur in varying proportions and the soils vary in the same relation. Its composition varies from almost pure sand to stiff red clay. It is usually poor in lime, iron oxide and nitrogen but fairly rich in potash some areas being quite rich in potassium derived from muscovite and feldspar present in the gneiss. Near the crest of the ridges the top soil is 3 to 5m thick while in the lower slopes it decreases.

Drainage

The rivers of the area drain ultimately to the south, though the west to east ridge across, it causes a series of Tista tributaries rising on its northern face to flow northwards and other flow east to west to joining the main river.

Its principal tributaries are Rilli Khola, Giel Khola and Peshok Khala on its left blank, the Ziyang and the Sevok on the right bank. The Tista is a broad mountain torrent with numerous shallows and rapids. Its current is swift and dangerous running in places at 14 miles/hr. and carried out different erosional processes like corrasion, corrosion, cavitations etc.

Its drainage and channel pattern changes with the structural configuration of the area. In the study area, trellis (sub-trellis, directional trellis, joint trellis) and dendritic pattern is common. So, it is clear from these drainage patterns that the underlying materials are uniform in nature in the entire area. The main stream *i.e.*, Tista is flowing sometimes straight, meandering and braided in its course.

Natural Vegetation

Vegetation can be classified into two broad groups; a tropical zone from the plains up to about 6000 feet, and a temperate zone from about 6000 to 12000 feet.

The lower part of the tropical zone comprising the slopes leading up to the base of the outer hills is characterised, especially in its western half, by forests of sal (*Shorea robusta*). The other valuable species in order of importance are auli lampate, panisaj, the rubber etc. These are mainly found along the valley side, but if we go away from the valley we will get a mixed type of forests. The most important species are safed champ (*Mihalia excelsa*), the timber of which is used for panelling the flooring in houses and is consequently in great demand, the tun (*Cedrela toons*) now used in house building, the buk (*Quercus lemellosa*), used for heavy beams and fuel, the singari catus (*Quercus pckyphylla*), the Pipli (*Bucklandia populnea*), and the Oak (*Juglans regia*).

Another important aspect in this area is the tea plantation. Tea plantation is carried out on the slopes which are low in nature and stable than the steeper one.

Mass Movement: Causes and Consequences

This chapter embodies the studies of mass movements and landslide hazard zonation mapping in the study area. Classification of various mass movements was undertaken and their relationship with various terrain factors was established.

Mass Movement Studies

Slope failures in the study area are mainly due to an accelerated erosion induced by an increased population pressure on limited land resources. The combination of adverse factors, *viz.* high relief topography, complexly folded rock formation and high rainfall make the problem of landslides and soil erosion too serious.

The masswasting in this area can be classified as surface erosion and mass movements as described below.

Surface Erosion

It is the removal of surface layer of the soil, also known as sheet erosion. Generally, the loss of top soil is direct consequence of rain drops pounding on the unprotected ground. In this area, it appears that sheet erosion is less, visually apparent than other forms of mass wasting but it is more damaging to the livelihood of the people. Sheet erosion is mostly manifested in the lower part of the study area.

Gully erosion by streamlets are also responsible for surface erosion. Usually associated with weak, jointed bed rocks, gullies mainly develop on the lower parts of tile valley sides slopes, where the protective vegetation cover is broken by sheet erosion or direct anthropogenic activities.

Mass Movements

It is a part of denudation process. Mass movement, which is an exogenic process of erosion, may be divided into slope process and subsidence. Crozier (1986) included the denudation among the four major groups of the rock cycle as given below:

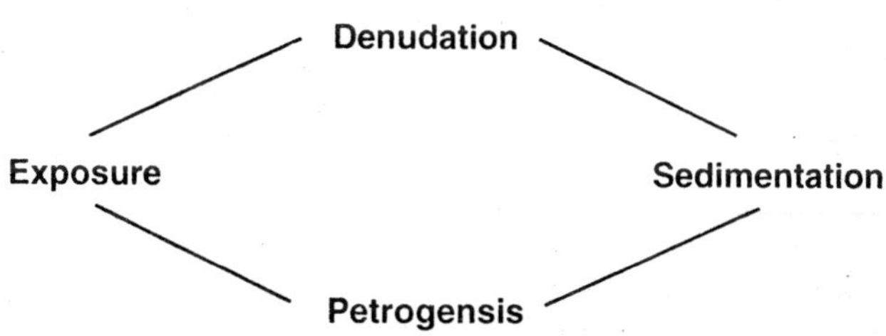

Figure 16.1: Rock Cycle Process

Classification of Mass Movements

The classification of the slope processes are known by various terms *viz.*, slope failures, mass wasting, mass movement and landslide.

The term landslide is used as a collective term for most of the mass movements excluding creep and flow. Skempton and Hutchinson (1969) described that the genetic term "landslide" embraces, slope movement of solid and rock masses resulted on the shear failure at boundaries of the moving mass. Varnes classified landslide on the basis of the types of movement and material. Types of movement are classified into 5 major groups such as *falls, topples, slides, spreads* and *flows*. The sixth group, *complex slope movement* incorporates the combination of two or more of the other type of movement.

Table 16.3: Classification of Landslides (Varnes, 1978)

Types of Movement		*Types of Material*		
		Bedrock	Engineering Soil	
			Predominantly coarse	Predominantly Fine
	Falls	Rock Fall	Debris fall	Earth Fall
	Topples	Rock topple	Debris topple	Earth topple
Slide	Rotation, Few units	Rock slump	Debris Slump	Earth Block slide
Slide		Rock block slide	Debris block slide	Earth Block slide
Slide	Translation, many units	Rock Slide	Debris slide	Earth slide
	Lateral Spread	Rock spread	Debris spread	Earth spread
	Flows	Rock flow (deep creep)	Debris flow (Soil creep)	Earth flow
	Complex	Combination of two or more principal types of movement		

The second classification was put forward by Crozier (1973), based on the morphometric characteristics of landslides, *viz.*, depth, dilation, tenuity, flowage, displacement and fluidity. The classification is given in Table 16.4.

Types of Mass Movements in the Study Area

The mass movements of the study area may be classified into following categories: mud flow associated with surface channel development; slumps or mud flow, debris or rock slide relating to percolation and landslide or slump created by sinking or lateral erosion etc.

Table 16.4: Morphometric Classification of Landslides (Crozier, 1973)

Process Group	*Class of Movement*
Fluid flow	Mudflows, debris flows, debris avalanches
Viscous-flow	Earth-flows, bouldery earth flows
Slide-flow	Slump/flow
Planar-slide	Turf-glide, debris slides and rock slides
Rotational-slide	Earth and rock slumps

Among the above categories, landslides are frequently noticed. Except these, there have also evidences of debris slide or rock slide and rock fall in the study area.

Debris slides are generally noticed on unconsolidated material at surface, due to saturation by water. Commonly debris slides are readjustment of slopes, and occur mostly, where internal resistance to shear is reduced below the safety limit.

Similarly *rock fall* is also a type of landslide, where blocks of rock variable in size may suddenly crash downward from the steep slopes. Such falls are generally observed along the road due to excessive cutting.

Slumps are of small magnitude *viz.*, debris slide or slumps, are generally observed on the valley side slope. If we go towards north, *i.e.*, Sikkim, the area between Sevokbazar to Tistabazar is considered as highly land slide-prone area, where we can observe all these hazards.

Mechanism of Landslides in the Study Area

Occurrence of landslides includes a number of mechanisms. A series of natural process along with human interference lead to landslide, but normally it is caused by trigger mechanism or threshold. There are many parts of Darjeeling Himalaya, which are noted for landslide, but they never experience landslide. Trigger mechanism is very much similar to pulling the trigger of a gun. But, it requires a long time plan.

To analyse it, we need to be well informed with the term stress and strain of the material. Stress is the force applied to a body, while strain is the resulting movement, of the body as a whole or its parts. Now, when the shearing stress is higher than the resistance or shear strain then a new process began.

Table 16.5: Relation Among Incidents of Landslides and Different Physical Parameters

Rocks	*No. of Slides*	*Slope Angle and Length*	*No. of Slides*	*Nature of Land*	*No. of Slides*	*Nature of Mountain*	*No. of Slides*
Sandstone	12	>50° >500 m	30	Rocky Wasteland	–	Mountain Ridge	2
Shale	7	>50° >500 m	15	Wasteland with Coarse Rock	5	Mountain Flank (Upper)	27
Phyllite	39	40°–50° >500 m	53	Converted Grazing Ground	16	Mountain Flank (Middle)	16
Schists	41	40°–50° <600 m	22	Grazing ground	3	Mountain Flank (Lower)	6
Slate	13	30°–40° >500 m	3	Tea garden	25	Valley (Upper)	8
Felsperic Rocks	6	30°–40° >500 m	15	Cultivable Land	19	Valley (middle)	19
Granite & Gneiss	24	20°–30° >500 m	14	Urban Areas	14	Valley (Lower)	26
Compound	26	20°–30° <500 m	8	Rurban areas	15	River bank Areas	76
Unknown	28	10°–20° >500 m	6	Rural areas	19	Compound	17
		10°–20° <500 m	3	Built-up areas	73		
		<500 m	2				

When there is huge downpours, it may trigger the process by exceeding the threshold value. It is remember that earthquake, mining activities, lightning, thunderstorm also act as trigger mechanism, but these are very rare cases.

The solid rocks are decomposed and formed regolith on hillslope. They are loose, porous and impervious upto the limit of parent rock. When heavy rainwater goes downward by infiltration and reach upto the contact zone with unaffected parent material; there will develop a temporary ground water table, technically known as "Piezometric head". Under the situation of normal rainfall, this head lies under the normal level, and thus hydrological balance is maintained.

But, in case of high rainfall, there is more water available for infiltration. So, temporarily ground water table moves upwards. These diverse movement of two water front, *i.e.*, coming up and going down displace the mass between these two water front and move downward, causes landslides. Perhaps it is the cause to occur landslides in hillslope.

Factors Affecting Landslides

Factors influencing landslides can be divided into two sets, permanent and variable. The major factors responsible for causing landslides in this study area are listed below:

Landslide in Relation to Slope

The influence of slope steepness on landslide occurrence is the easiest factor to understand. Generally, steeper slopes have a greater chance of land sliding. This does not prevent failures from occurring on gentler slopes. Other factors may make a gentle slope especially sensitive to failure and thus in this, situation it could be determined to have a relatively high hazard potential. Conversely steep slopes may not always be the most hazardous.

In the study area, average slope values determined by Wentworth's method has yielded 5 classes from less than 15° to more than 30°. It has been classified from gentle slopes (<15°) to very steep slope (>30°). The highest frequency of landslides occur in the group >30° and followed by 25° to 30°. Mainly along the NH 31 in these classes we generally observed this landslides phenomena.

Landslides in Relation to Relative Relief

On the basis of altitudes the area can be divided 6 relief zones. The three relative relief zones *viz.* 1200 to 1800 ft., 1800 to 2400 m and 2400 to 3400 m have experienced maximum landslides. The amount of landslides gradually decreases towards lower and higher relief zones.

Landslides in Relation to Hydrology

In the area under study overland flow was a common phenomena during rainstorms. This may have been due to precipitation intensities exceeding soil infiltration with consequent surface runoff over unsaturated soils. The length of the rainy period and intensity and duration of a single rainfall event makes piezometric surface even above the ground surface. The effectiveness of piezometric levels maximize the stable slope saturation and cohesionless soil mantle with downslope seepage is higher than unsaturated mantle (O'loughlin, 1972).

Changes in river courses modify perennial water resources and change in surface and ground water hydrology create problems throughout the Himalayas. On the other, the constant moisture reduces cohesive strength and lubricate failures planes. Such type of process could be embraced in the continuous actions of the slow debris flows in the drainage basins during the rainy and summer months.

Landslides in Relation to Anthropogenic Activities

Despite the natural factors causing mass-wasting hazards, practice of anthropogenic technology has been capable to simulate drastic changes in physical environment. Such changes can be divided into two categories:

(*i*) changes in landuse by removal of forest cover and other exploitation, and (*ii*) Construction works, principally roads rail, dams, Hydro-electric power stations and other infrastructure activities.

In the study area, removal of forest for many purposes have initiated increasing rates of soil to deterioration, reducing the moisture holding capacity and as a result, increasing the incidence of soil erosion, land-sliding and gullying. People who live there need and exploit local resources continuously for constructing houses, walls, roofs. They also carry stones and soils which cause the loss of the basal support of the surface. Thus causing, the landslides hazard.

Construction works like dams, hydro-electric power stations, bridges, roads and other activities have fostered the mass wasting in the study area. There, the intensity of the vehicles playing in this road (NH-31) is increasing day by day and as a result it not only reduce the slope stability but also the environmental conditions of that area. Similarly poor alignments, poor construction technique has led to destabilize slopes through the removal of toe, support and abutment.

Mitigation and Stabilisation of Mass Wasting Hazards

Mass wasting hazards are the outcome of a number of physical, cultural and social activities. There are several mitigating and stabilization measures of hazards for which measures like proper identification and the preparedness against the occurrences are sought.

Varne's (1984) definition and differentiation are generally accepted, such as the definition for Natural Hazards (NH): the probability of occurrence of a potentially damaging phenomena within a specified period of time and given area; the definition of Vulnerability (V): the degrees of lost to a given element or set of elements at risk resulting from the occurrence of a given magnitude which is expressed on a scale from 0 (no damage) to 1 (total loss); the definition of risks (Rs): the expected degree of loss due to a particular natural phenomena. It may be expressed by the product of H and V with element at risk (E): the population, properties and economic activities including public services etc. at risk in the given area and that of total risk (Rt): the expected member of lives lost, persons injured damage to property, or discretion of economic activities including public services etc at risk in the given area and that of total risk (Rt): the expected number of lives lost, persons injured, damage to property, or disruption of economic activity due to particular natural phenomena. It is, therefore, the product of specific risk (Rs) and Element at risk (E), *i.e.*, Rt = (E)* (Rs), (Rs) = (H*V).

Imperatives of Mitigation

There is an important question arise whether to mitigate a landslide or a landside prone areas. In our country, only mitigating a landslide is practiced.

As it is a natural process, which occurs and recurs, in certain geological settings under certain conditions. The rising costs of landslide damages are a direct consequence of the increasing vulnerability of people and structures to the hazard. Therefore, in. order to identify and address the perspectives and concerns of the various communities, the groups affected by landslide hazards, who are in acute situation, need to be consulted immediately. The vulnerability of people to natural hazards is determined by the relationships between the occurrences of extreme events, the proximity of people to these occurrence and the degree to which the people are prepared to cope with these extremes of nature.

Map Analysis

It is usually one of the first steps in a landslide investigation. Maps that can be used for this purpose include geologic, topographic, soil and geomorphic-maps. In the study, the variables like, slopes, reliefs, drainage, population have been used intensively to study the landslide hazard in the study area.

Analysis of Aerial Photography and Imagery

It is a quick and valuable technique for identifying landslides because it provides a three-dimensional over view of the terrain and indicates various land-use patterns with geological information. The phases of activities for landslide hazard are shown below:

Computerized Landslide Terrain Analysis

This technique helps us to determine the volume of landslide masses and changes in surface expression and cross section over time. This information is useful in calculating the potential for stream blockage, cost of landslide removal and type and mechanism of movement. Very promising methods are being developed utilising Digital Elevation Models (OEMs) to evaluate areas quickly for their susceptibility to landslide/debris flow system.

Instrumentation

Modern instrument such as Electronic Distance Measurement (EDM) and others like inclinometer, extensometer, strain meter, tilt meter and piezometer are used to find out the mechanics of landslide movement and to warn against slope failure.

Importance of Mapping

Maps are a useful and convenient tool for presenting information on landslide hazards. They can present many kinds and

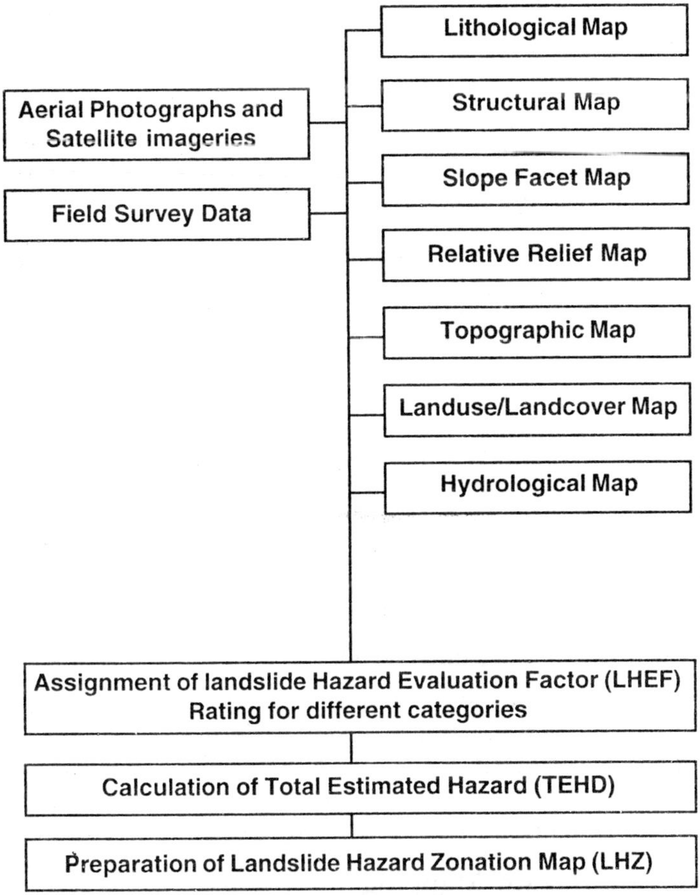

Figure 16.2: Phases of Activities for Landslide Hazard Zonation

combinations of information at different levels of details. Leighton (1976) suggested a three stage approach to landslide hazard mapping: the first stage is regional or reconnaissance map the second community level mapping and the third site specific.

Community-level Mapping

This type of mapping identify both the three-dimensional limits of landslides and their causes. Guidance concerning land use, zoning and building as well as recommendation for future site-specific investigations are also made at this stage.

Site-specific Mapping

It is concerned with the identification, analysis and solution of actual site specific problems. It is usually undertaken by private consultants for site development and for procuring necessary information for design and construction. Though the present study has used basically regional inventory type of map, it has however, depicted and apparent glance of site specific problems.

Landslide Hazard Mapping and the Development Planning Process

The susceptibility of a given area to landslides can be determined and depicted using hazard zonation. It can be used as a tool to help identify land areas best suited for development by examining the potential risk of land sliding. As a result, landslide hazard is often represented by landslide susceptibility (Brabb, 1985).

Even with detailed investigation and monitoring, it is extremely difficult to predict landslide hazards in absolute terms. Sufficient understanding of landside processes does exist, however, to be able to make an estimation of landslide hazard potential.

Determining Acceptable Risk

Natural and as well as human-induced changes can affect the susceptibility to landslides and must be understood when assessing the landslide potential of an area. A decision is ultimately made regarding the degree of risk that is acceptable or unacceptable to a project. Mitigation strategies are then designed to reduce risk.

Landslide Hazard Mapping

To understand landslide hazard in future we must go through in detail with the bedrock, past history, slope steepness of the study area. Specific combinations of these factors are associated with differing degrees of landslide hazards. The identification of the extension of these combinations over the area being assessed in a landslide hazard map.

Development Planning Process or Integrating Landslide Hazard Zonation Maps

It is important to identify relative landslide hazard levels early in the planning process. This permits planners to determine the degree of landslide risk that is acceptable or unacceptable to a development program. It is carried by the following steps:

Preliminary Mission

An initial review is made of the type and content of available information, including natural hazard information. During this stage of study available information should be collected and reviewed concerning assessments of natural hazards, including landslides and disasters, which are known to have affected the study area.

Development Diagnosis

Here we identify the areas that have highest development potential, known as target areas. It involves identifying and delineating natural resource factor that favour or constrain development in a particular area. Landslide hazard is an undesirable factor, and the greater the hazard, the more it may shape the development potential.

When a potential hazard is present in the study area, the first step is to undertake a brief survey to establish whether landslides have occurred in recent times or not. Roads, railroads and river banks are good sites for seeking signs of past landslide occurrence. However, activities may increase landslide hazards and the absence of evidence from past landslides does not guarantee that landslides will not pose any problem in the future.

The degree of landslide hazard in an area is a limiting factor only for those activities that may alter the existing balance between forces driving and resisting movement on an unfailed slope. Planners need to understand what effects development activities may have on this balance of forces. For example, removing forest cover to create a field for cultivation crops is much more likely to lead to landslide occurrence, since it alters the balance of forces and may increase the susceptibility to slope failure by some triggering event, such as prolonged rainfall, which would not have produced the landslide under the original conditions.

Development Strategy and Project Formulation

Development projects considered for the target area are formulated at this stage. Also at this time, landslide hazard evaluation within the study area is refined. Mitigation measures should be selected if they are not already part of the project identification formation. Thus, it is possible to reduce the possible impact of natural landslide activity and limit landslides which occur as a result of human activity.

Project Implementation

A detailed hazard map for the specific site is necessary at this stage. The large-scale features represented on landslides mapped in this detailed inventory are valuable for test drilling at a site and other sampling activities of engineering design work. As well as different hazards areas are targeted for development with greater detail about the hazard potential.

Preventing and Stabilizing Landslides

Hazardous landslide-prone slopy location and activities can be stabilised by proper identification and preparedness against such occurrence. The methods normally used to protect landslide are as shown in Figure 16.3.

Retaining Wall

It is the mast common method, which provides support along the cut slope and often erect to offer additional basal support along the cut slope. In Darjeeling Himalaya mainly this method is applied to lessen the intensity of landslide.

Construction and Maintaining a Proper Drainage

Landslide is more or less related with rainfall and drainage parameters (both surface and sub-surface). It not only reduce the possibility of landslide, but also helps in terrace cultivation.

Sheet Pilling

It is a expensive method in which the hillslope is covered either by aluminum or steel sheet. But it is not practiced in this study area.

Rock Boll and Anchoring

This method is not yet used in the Darjeeling Himalaya. But it is already in action in Joshimath to Badrinath area where rockslides are common phenomena. These are operated by clamping the unstable rock block back to the stable rock beneath. The force exerted by tension on the anchor opposes the force that tends to push the rock block down the slope. Anchors must be installed at inclination over than 900 to the dip of the plane of sliding.

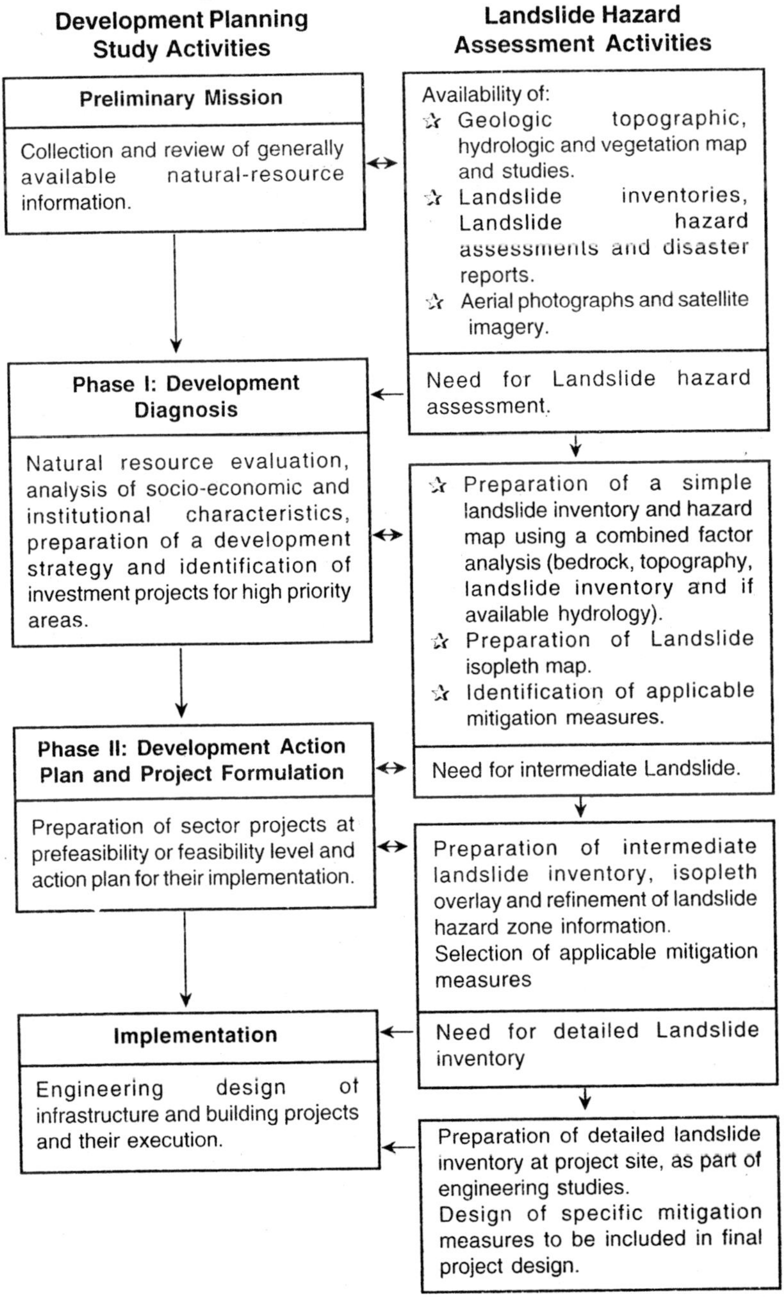
Development Planning Study Activities
Landslide Hazard Assessment Activities
Preliminary Mission
Collection and review of generally available natural-resource information.
Availability of:
☆ Geologic topographic, hydrologic and vegetation map and studies.
☆ Landslide inventories, Landslide hazard assessments and disaster reports.
☆ Aerial photographs and satellite imagery.
Need for Landslide hazard assessment.
Phase I: Development Diagnosis
Natural resource evaluation, analysis of socio-economic and institutional characteristics, preparation of a development strategy and identification of investment projects for high priority areas.
☆ Preparation of a simple landslide inventory and hazard map using a combined factor analysis (bedrock, topography, landslide inventory and if available hydrology).
☆ Preparation of Landslide isopleth map.
☆ Identification of applicable mitigation measures.
Need for intermediate Landslide.
Phase II: Development Action Plan and Project Formulation
Preparation of sector projects at prefeasibility or feasibility level and action plan for their implementation.
Preparation of intermediate landslide inventory, isopleth overlay and refinement of landslide hazard zone information.
Selection of applicable mitigation measures
Need for detailed Landslide inventory
Implementation
Engineering design of infrastructure and building projects and their execution.
Preparation of detailed landslide inventory at project site, as part of engineering studies.
Design of specific mitigation measures to be included in final project design.

Figure 16.3

Table 16.6: A Hundred Years of Landslide in Darjeeling Hills

Date of Slide and Rainfall	*Affected Areas*	*Remarks*
24–25 September, 1899 1065.5mm	Darjeeling Town (Toongsong, Alubari, Rail station, Jalapahar) Kurseong, Kalimpong, Ghoom, Tindharia, Bloomfield and Takdor Garden	72 people died; many buildings collapsed.
15 January, 1934 Bengal–Bihar earthquake	Darjeeling Town, Sonada, Sukiapokhri, Ghoom and Kurseong	Darjeeling Town and adjoining areas are mainly affected.
11–13 June, 1950 420mm	Darjeeling Town (Jalapahar, Lebong, Katapahar, Bhutia Basti) Happy Valley, Kalimpong, Mahanadi, Paglajhora, Tindharia, Kurseon Takdah	127 killed; Siliguri-Kalimpong rail route collapsed the area was isolated for 5 days
2–6 October, 1968 1121.4mm	Darjeeling Town (Toonsong, Kotwali, Kagjhora, Manpari, Lebong) Kalimpong, Teesta Bazar, Rambi, Algarah, Peshok.	Hill–Cart road and NH–31 were totally abolished; Many people died.
3–5 September, 1980 299.1mm	Rimbik, Lodhama, Bijanbari, Darjeeling Town, Sukiapokhri, Ghoom, Manebhanjan, Sonada, Lebong, Tindharia, Happy Valley Ambutia	215 people died in Rimbik and Lodhama; Approximate loss of property was 100 minion rupees
15–16 September, 1991 462.5mm	Darjeeling Town (North Point, Toongsong, Singamari), Takdor, Bloomfield, Rongtong, Pagalajhora, Tindharia, Chunabhati	2 people died; Siliguri Darjeeling rail route blocked for 5 months.
11–13 July, 1993 211.3mm	Mongpo, Takdah, Peshok, Rongtong, Tindharia, Pankhabari, Mahanadi, Gavabari, Ambutia, Darjeeling Town	15 people died in Mongpo
June, 1997 301–475mm	Gangtok, Paglajhora, Darjeeling Town, Rongtong, Tindharia, Toong, Rangpo) Singtam	Loss of property worth Rs. 1.5 billion; 150 people died.
July, 1997 312–487m	Teesta Valley, Mongpo, Gangtok, Eastern and Northern Sikkim	67 people died; Siliguri-Gangtok road was very much affected.

Hardening of Soil

This is used by National Hydel Power Corporation (NHPC). It is also very expensive method, in which, electricity is passing through soil to transform the soil into rock. Actually, it is the method of recrystallisation of soil at 3000–3500°C and it requires 24 hours time to recrystallise. But in the study area it is not been used by the govt. to stable the area.

Afforestation

The beneficial role of natural vegetation against landslide hazard has been demonstrated in many parts. Logically, in catchement area it may solve the problem to a large extent. However monoculture of certain coniferous species like Dhupi is found not so effective. It is suggested that deciduous bushy grass are highly effective against slope failure.

Restriction to Settlement

As population plays a vital role in degrading the environment. So, there should be restrictions for further encroachment which must strictly be adhere to slope. Beyond 30° they are not allowed for their settlement and also the area should be free from agricultural practices.

Terracing and Agricultural Practices

Terrace cultivation must be followed with contour line along with proper drainage outlets. Many parts of the Darjeeling Himalaya this technique is followed to stabilize the slope and it is a really give us a good result.

Discussion and Conclusion

The main objective of the present study is to study the landslide problem in the area between Kalijhora and Tista Bazar along Tista Valley and to establish the relationship between various morphometric parameters.

Investigations reveal the following points:

1. Morphogenetically this area is divisible into three units, summit surfaces, hill side slopes and valleys with their terraces.

(*a*) Rock fall is the dominant process on free faces, whereas landslides and other types of slope failures are frequent in the lower parts.

(*b*) Valleys have gentle aggradational slope where gully erosion appears to be the dominant degradational process.

(*c*) The meanderings of Tista is due some lithostructural control.

2. Various morphometric parameters of linear, areal and relief aspects of the drainage basin were analysed.

(*a*) The linear aspects show the sincerity and meandering nature of the camels.

(*b*) Aerial aspects indicate that larger the basins have higher discharge and low infiltration. In this way, the stream frequency maintains a close relationship with drainage density.

(*c*) The relief aspect shows that the entire area is in youthful stage with a high erosive activity.

(*d*) The high angle of slopes and high relative relief indicate higher rate of this hazard.

3. Undercutting of slope on fluvial action, deforestation, unscientific constructional activities etc. are responsible for bewildering varieties of slope failures. It was observed that surface erosion is a dominant process of mass warding on the valley side slopes.

(*a*) Three categories of mass movement like debris slide rock fall and landslide observed in the area.

(*b*) The analysis of the factors like inherent, acquired and external gives all idea about the possible causative factions. The inherent factors are lithological and structural status of the bed rocks and hydrological conditions, whereas the acquired factions are indiscriminate road cutting, increase of head load etc. white the external factors may be seizure city and very high rainfall.

Management and Planning in the Area

The natural problems like landslide hazard are the result of the combined effect of various factors. Some of the aspects and measures that should be taken into consideration in the future are presented as follow:

1. The objectives, policies and programmer concerning environmental conservation, landuse and sustainable development should be taken into consideration and effectively implemented.
2. Deforestation of the natural vegetation should be checked.
3. Hilly areas are rich in natural resources, and their exploitation is a common phenomena there. So, to prevent slope failure, resources should be utilized in such a way that the nature is not disturbed.
4. Transportation rules should be replanted. New settlements are growing up and as result, different road networks are developed which not only hamper the slope stability but also disturbed the maintain ecosystem.
5. Mineral resourses should not be extracted in order to prevent this landslide hazard as it degrade the slope stability and increase the intensity of the landslide by fracturing and cracking in the rock surface.
6. Settlement should be restricted. It settlement is developed without any plan, then the area will be immediately gone under some natural problem like water scarcity deportation of natural resources and slope failure.
7. High rise building should not be allowed to grow along major access routes as its pressure on the soil reduce the slope stability.

References

Bloom, A.L., 1991. *Geomorphology: A Systematic Analysis of Cenozoic Landforms*. Prentice Hall, Engle, Woodcliffs)

Carson, M.A. and Kirkby, M.S., 1972. *Hill Slope, Form and Process.* Cambridge University Press, London, p. 112--115.

Chorley, R.J., 1969. The drainage basin as a fundamental geomorphic unit. In: *Introduction to Physical Hydrology,* (Ed.) R.J. Chorley. Mehtuen and Co. Ltd., London, pp. 30–52.

Doornkamp, J.C. and King, C.A.M., 1971. *Numerical Analysis in Geomorphology*. Edward Arnolds, London, pp. 1–181.

Leopold, L.B., Woman, M.G. and Miller, J.P., 1964. *Fluvial Processes in Geomorphology*. W.H. Freeman and Co., USA.

Mukhopadhyay, S.C., 1982. *The Tista Basin: A Study in Geomorphology*. Ram Krishna Printing Works, Calcutta.

Strahler, A.N., 1956. Quantitative slope analysis. *Bull. Geol. Soc. Amer.*, 67: 571–596.

Singh, Savindra, 2003. *Geomorphology*. Prayag Pustak Bhawan, Allahabad, pp. 353–384.

Young, A., 1963. *Deductive Models of Evolution of Slopes*, Comm., Dep., 3: 45–66.

Chapter 17

Land Use Utilisation in Darjeeling Himalaya

Debashis Das

ABSTRACT

The article deals with a study of Land Use in Darjeeling District in the Eastern Himalayas. An attempt has been made to scientifically and methodically classify the present land use in the areas of the study. The parameters kept in mind were the spatial distribution of different categories of land use utilization, their inter-relationship and human activities in the villages with different population sizes. The results have shown that forest resources are concentrated in the elevated North-Western part of the district. This part is not available for cultivation. In moderately elevated areas the forest resources are not so concentrated and there are no forests in the plains but cultivable and wastelands are available.

Introduction

Land use classification is the systematic arrangement of various classes of land on the basis of certain similar characteristics, mainly to identify and understand their fundamental utilities in satisfying the needs of human society. The best use of each parcel of land requires a scientific and methodically appreciable classification of the present land use. This may help us in investigating the land use problems and be the basis of planning for the best use of our land after considering the major land use categories (Mandal, 1982).

The study of cropping pattern is a scientific device to analyse the spatial relationship of crops in association with each other in land utilisation. The idea of cropping pattern could be best utilised in understanding the existing agricultural situation, the land use pattern of a region and planning of crops for the future (Mandal, 1982).

Objectives

This study has following objectives:

1. To analyse the spatial distribution of different categories of land utilisation.
2. To study the interrelationship between different categories or land utilisation.
3. To examine how the lands are used under different activities in the villages with different size classes of population of the areas under different P.S. of the district.

Location and Physical Set Up of Darjeeling Himalaya Region

Darjeeling Himalaya Region is located at northern part or the state and eastern part of the country. This is a mountainous region. It lies between 87°57′ E and 87°59′ E longitude and 26°48′ N and 27°13′30″ N latitude. Total area of the district is 325.4 thousand hectares, which is only 3.68 per cent of West Bengal. The district includes three hill sub-divisions, (*a*) Darjeeling Sadar, (*b*) Kurseong, and (*c*) Kalimpong and one in the plain, *i.e.*, Siliguri sub-division.

Darjeeling division is traversed by a long ridge entering the area from the north. The average elevation of this range varies between 1875 m and 2000 m. The elevation of the area where tea estates are found ranges between 625 m and 1250 m. Cultivation of most of the crops takes place in that part of the region. This region is drained by the Tista and the Great Rangeet rivers. The Kalimpong division has a broad mass or a high land in its north central part. The height of this part ranges between 156 m and 3152 m. This subregion is covered by thick forest. It is drained by the Tista and its tributaries. This part has less drainage density (Lahiri, 1981).

Geologically this region can be sub-divided into three zones: (*a*) an outer belt or Siwalik frontal range, (*b*) a medium narrow belt or Damuda Range, (*c*) and an inner range of the metamorphites.

Soils derived from this group of rocks are porous sandy loam often containing humus except in rocky areas. These soils are grouped under grey brown to forest soil with great depth in certain areas. Red soils are also found in many areas. These are due to their iron content which are derived from the ferro-magnesium silicate. More than 94.5 per cent of the sandy loam soils are acidic and rest 5.5 per cent are neutral. Among the plant nutrients, nitrogen is available in sufficient quantity and phosphorus and potassium are available in moderate quantity (Lahiri, 1981).

Temperature below freezing point is felt every year in higher areas. Snow falls occasionally above 2187 m. Frosts disintegrate rocks and form soils. Melting of snow and spring water irrigate crops in the dry reason. The region gets heavy south west monsoon rainfall from June to September. The mean annual rainfall exceeds 610 cm on the north central range of Kalimpong. The southern part of the ridge gets more than 350 cm of rainfall but on the top and on the lower slopes it decreases.

Spatial Distribution of Land under Different Uses

Forest is an important economic resource of West Bengal. It is well adapted on the steep slope of the terrain, in heavy rainfall and on thin soil. Forest coverages predominate over 312 m of height. Due to the variation of terrain character, rainfall, soil and biotic factors the species and composition of trees change considerably from place to place. As for example Sal trees grow extensively on the southern slope of mountains, whereas miscellaneous trees predominate on the northern slope. Low rainfall is available at the river Rangeet Valley where Chir pine trees are found, but these trees are not found in other areas.

Forest divisions are formulated at Darjeeling, Kurseong and Kalimpong areas for better management of forest resources. The objectives of these forest divisions are as follows:

1. To retain a forest cover for preventing erosion and landslides during heavy rainfall.
2. To assist in the conservation of water supply.

3. To convert wherever possible the present irregular mixed and often aged forests with defective trees into regular forests with useful species of trees.
4. To exploit the existing crops to the best advantage, so that highest monetary yield could be obtained.
5. To maintain and tend the regenerated crops.
6. To meet the local market demands of timber and other forest produce (Lahiri, 1981).

Although, these forests are managed by the department of forests of West Bengal government, yet, commercial forestry has not developed here. Another project is set-up to look after the developmental programmes of the forests. This project includes the programme of development of all weather roads, mechanisation of logging operations, log hauling by rope ways, afforestation of felled areas, reforestation of all areas from where timber is harvested and finding out markets for blank and burnt areas, forest produce by establishing or helping to establish wood based industries.

Commercial forestry depends on the accessibility of the forests. Out of 82,362 hectares of forest land 36,862 hectares or 44 per cent are accessible, 12,500 hectares or 16 per cent are partially accessible and rest 33,000 hectares or 40 per cent are inaccessible.

Some of the natural forests in the accessible areas have already been converted into planted forests. The remaining natural forests are situated on difficult terrain. Often in higher areas the trees have been found to be mature to over mature and quite a good percentage is found to be defective.

It is observed from the statistical abstract of West Bengal, 1975 that there are 118.3 thousand hectares of land or 38.07 per cent is devoted under forest in Darjeeling district since 1960–61, which remained the same in other years also upto 1967–68. According to the "Estimates of area and production of principal crops in West Bengal, Directorate of Economics and Statistics," 33.9 thousand, hectares of land or 10.41 per cent were under forests in Darjeeling in 1984–85. But, according to district census handbook, Darjeeling district, 1971, the percentage of forest land was 20.45 per cent.

Forest resources are extensively found in the north western part of Darjeeling district. This part includes areas under Sukhiapokri,

Pulbazar, Rangli Rangliat, Jore Bunglow and Darjeeling P.S. where the proportion of forest areas ranges between 31.47 per cent and 68.47 per cent. Forest coverage decreases towards the Southern part of the district. The density of forest also decreases from north west to central and north eastern part of the district. The north western part has 7,33,398 acres or 87.09 per cent of the total forest area of the district, that means rest areas under other (eight) P.S. have only 12.91 per cent of the districts' total forest areas. This indicates high concentration of forest resources in a limited area. This area is very much undulating where the elevation ranges between 1000 metre and 7000 m. The elevation and ruggedness diminish towards other parts of the district, where deforestation predominates. Hill soils are found there.

Cultivable Wasteland

These lands are cultivable, but due to floods, landslides and other reasons the quality or land becomes degraded, the soil becomes unproductive. These are the wastelands. In this district 42,934 acre of land (10.42 per cent) are cultivable wastes, which are mainly found in the areas under Kharibari, Phansidewa, Kurseong, Mirik and Kalimpong P.S. where the proportion or wasteland ranges between 13.66 per cent and 25.51 per cent. These areas together concentrates 30419 acres or 70.85 per cent the total wasteland or the district. This indicates that waste land is concentrated in the southern, central and eastern part of the district, where proportion of forest covered land is quite low. So, due to deforestations and landslides, soil erosion predominates there which increases the amount of wasteland significantly.

Area Not Available for Cultivation

This area includes the rugged and undulating part of the district where agriculture is impossible and it also includes those lands where residential areas are found or where different infrastructure and facilities are located. The district as a whole has 62,228 acres areas that are not available for cultivation, which is 15.09 per cent of the total area. This kind of land is extensively found in the north eastern and eastern part of the district, that includes the area under Garubathan, Kalimpong, Kurseong and Siliguri P.S. The proportion of area not available for cultivation ranges between 18.31 per cent and 29.03 per cent in the above mentioned areas, which together constitute 39,873 acre of land or 64.07 per cent of the total area that is not available for cultivation of the district. The entire area has the elevation ranging between 500 and 3000 meter, which is lesser than the north west of western part of the district. This part has more or less homogeneous resources, where large number of settlements, including three out of four urban centres, are located. Besides most of the economic activities are performed there, where diversified occupations and large commercial nodes exist. All these prove that infrastructure, facilities, and amenities are largely concentrated in this part of the district.

Net Sown Area

Net sown area is the cultivated area or where cultivation takes place at least ones in a calendar year. This area may be found extensively on the plain terrain or in the valleys. The district as a whole has 2,22,725 acres or 54.04 per cent of the total land as net sown area. This percentage of the state is 62.91 per cent and of the country is 46.66 per cent. This means the mountainous district of West Bengal has small proportion of net sown area than the state located at the lower part of Ganga plain, but has larger proportion than the country average. Large proportion of net sown area is found in the southern and north-eastern part of the district. This area includes areas under Naxalbary, Siliguri, Phansidewa, and Garubathan P.S., where the proportion ranges between 63.36 per cent and 81.70 per cent. Besides, these areas together concentrate only 62,243 acres or 27.94 per cent of N.S.A. of the district. This means, unlike other aspects of landuse net sown area is more or less equally distributed in the areas under different P.S. of the district. The southern part is a low elevated area with plains, who's elevation ranges between 250 meter and 500 meter. This part is covered by new alluvium soil. On the other hand the height of north eastern part of the district is 3000 meter, which is covered by hill soils. It is interesting to note that the areas with high proportion of NSA have low proportion of net irrigated area, and vice-versa.

Relationship Between Different Elements of Landuse

The relationships between different elements of landuse are identified with the help of correlation coefficient technique.

The formula is:

$$r = \frac{\Sigma xy - \frac{\Sigma x . \Sigma y}{n}}{\sqrt{\Sigma x^2 - \frac{(\Sigma x)^2}{n}} \sqrt{\Sigma y^2 - \frac{(\Sigma y)^2}{n}}}$$

The formula of test of significance is:

$$t = r\sqrt{\frac{n-2}{1-r^2}}$$

The study reveals that proportion of area under forest is positively correlated with only the proportion of net irrigated area, but is negatively correlated with proportion of area not available for cultivation, proportion of net sown area and proportion of cultivable wasteland. This indicates that the forest resources are extensively available mainly on the north western elevated part, where vegetables and other crops are cultivation on the terraces with local irrigation schemes. Paddy, the rainfed crops, is not cultivated widely over that sub-region. These crops are cultivated on limited areas, in some pockets, only. Due to the persistence of huge forest areas, the net sown areas and areas not available for cultivation are not found extensively in the region.

Proportion of area not available for cultivation is positively correlated with proportion of net sown area and proportion of cultivable waste land, indicating higher the proportion of area not available for cultivation, higher are the proportions of net sown area and cultivable waste land and lower the proportion of forest land. Although proportion of area not available for cultivation and proportion of net irrigated area are positively correlated, but that is insignificant, means the relationship is the chance factor only. It may not show same relationship if larger observations are taken care of. Area not available for cultivation includes such areas which are used under settlements, under the construction of infrastructures, facilities and other welfare arrangements. That means these areas are suitable for the residence of people, and the people get enough employment at least in agriculture. Therefore, the areas have high

Table 17.1: Correlation Co-efficient Values

	Per cent of Forest	*Per Cent Area Not Available for Cultivation*	*Per cent Net Sown Area*	*Per cent Cultivable Wasteland*	*Per cent Net Irrigated Area*
Per cent of forest area	1.00	(–) 0.63	(–) 0.82	(–) 0.50	0.55
Per cent area not available for cultivation	(–) 0.63	1.00	0.49	0.35	0.09
Per cent net sown area	(–) 0.82	0.49	1.00	0.06	(–) 0.46
Per cent cultivable wasteland	(–) 0.50	0.35	0.06	1.00	(–) 0.08
Per cent net irrigated area	0.55	0.09	(–) 0.46	(–) 0.08	1.00

N.B. N. = 13

(–) 0.82: Significant at 0.01 level of significance.

(–) 0.63: Significant at 0.05 level of significance.

(–) 0.50, (–) 0.55, 0.49: Significant at 0.10 level of significance.

(–) 0.46, 0.35: Significant at 0.50 level of significance.

0.09, 0.08, 0.06: Insignificant.

proportion of net sown area. This area has small proportion of forest or vegetation covered area, so that the problems of soil erosion, landslides are acute. Moreover, these areas suffer from flood problem. All these facts cause high proportion of cultivable wasteland in the area. This area, although has high proportion of N.S.A., but the proportion of net irrigated area is small. That means agriculture is not assisted adequately by irrigation.

Proportion of net sown area is positively correlated with proportion of area not available for cultivation, but negatively correlated with the proportion of area under forest and net irrigated area. The proportion of net sown area and proportion of cultivable wasteland area although positively correlated, but the relationship is insignificant. That means it is chance factor only, it may not hold good in case of large number of observations.

The proportion of cultivable waste land is positively correlated with proportion of area not available for cultivation, but negatively correlated with the proportion of forest land. The reasons are already stated previously. The proportion of cultivable wasteland is positively correlated with proportion of net sown area and negatively correlated with the proportion of net irrigated area. But both these relationships are insignificant, indicating that their relationships are chance factors only.

The above analysis indicates two distinct subregions in the district. The north, western region is (5 P.S.) elevated, which is covered by hill soils. This undulating part is covered by forest resources to a great extent. Small patches of agricultural lands are found in some terraces where vegetables and other crops are grown with the help of irrigation. This region has small amount of area not available for cultivation, net sown area and cultivable wasteland.

Another subregion covers rest parts of the district, *i.e.*, southern, central, and north eastern part of the district. This subregion includes areas under rest eight police stations. This subregion has relatively low relief, which is covered by hill soils on the north east and new alluvium soils in the south. This part of the district has high amount of areas not available for cultivation, net sown area, cultivable waste land, but has small amount of forest and net irrigated area. This sub-region produces paddy and other crops without getting irrigation.

Land Utilisation in Villages According to Settlement with Different Size Classes of Population

The objective of this part of study is to find out that how the people of rural settlements with different size classes of population use the land. For this purpose out of thirteen areas under P.S. four areas have been chosen after considering elevation and land use characteristics. The height of areas under Garubathan P.S. ranges between 300 m and 5000 m. This area has highest percentages of area not available for cultivation. The height of areas under Kurseong P.S. ranges between 1000 m and 4000 m. This block has highest percentage of net irrigated area. The elevation of areas under Siliguri P.S. ranges between 50 m and 750 m. These areas have high proportion of net sown area. The elevation of areas under Kharibari P.S. is 250 m. It has highest percentage of cultivable waste land.

Area under forest is considerably large in the north western part of the district. In that part a small number of settlements are located at areas under individual P.S. where land is mostly under deep forests, where other uses of land are rarely found. That is why not a single area under P.S. is chosen from that part.

Out of four areas Garubathan (0.95 per cent) and Kurseong (15.52 per cent) P.S. have forest resources. In area under Garubathan P.S. 100 per cent of the land of very small sized settlements are under forest and only 0.40 per cent of the land is under forest in case of medium sized settlements (Table 17.2) and rest settlements not have forest land at all.

Table 17.2: Distribution of Forests (Percentage)

Size Classes of Population	*Name of Area under Police Station*			
	Garubathan	*Kurseong*	*Siliguri*	*Kharibari*
Less than 200 (very small)	100	0	–	–
200–499 (small)	–	1.58	–	–
500–999 (Low medium)	–	8.91	–	–
1000–1999 (medium)	0.40	20.41	–	–
More than 2000 (Large)	–	18.76	–	–

In areas under Kurseong P.S. medium and large settlements have considerable proportion of area under forest which decreases towards smaller sized settlements. It is therefore evident that these areas have opposite trend of having forest covered land in settlements with different size classes of population.

The percentages of cultivable wasteland are 6.63 per cent in Garubathan (3000 m–5000 m), 15.88 per cent in Kurseong (1000 m–4000 m) 10.16 per cent in Siliguri (500 m–750 m) and 25.51 per cent in Kharibari P.S. (250 m). This indicates that the elevation and proportion of cultivable wasteland are negatively correlated to each other.

The areas under Garubathan and Kurseong P.S. are located at relatively more elevated part of the district than area under Siliguri and Kharibari P.S. Therefore, in the more elevated part the size classes of settlements and percentage of cultivable waste land are inversely proportional to each other, whereas in plain land both are directly proportional to each other (Table 17.3).

Table 17.3: Distribution of Cultivable Wasteland (Per cent)

Size Classes of Population	*Name of Area under Police Station*			
	Garubathan	*Kurseong*	*Siliguri*	*Kharibari*
Less than 200 very small	–	31.78	8.56	20.77
200–499 small	30.33	6.65	9.25	23.43
500–999 low medium	7.77	29.45	11.84	25.96
1000–1999 medium	6.27	14.88	8.31	28.14
More than 2000 large	6.18	11.64	12.62	29.63

In the elevated area the small or very small sized, settlements of area under Kurseong and Garubathan P.S. have nearly on third of their land as cultivable waste, which decreases to 11.64 per cent and 6.18 per cent in case of large settlements. The cultivable wastelands are unequally distributed in these two sizes of settlements in the elevated part of the district. On the other hand, this kind of land is not to that extent unequally distributed in case of small and large size settlements, than are located in the plains, (Table 17.3).

The percentage of area not available for cultivation and size of settlements are directly proportional to each other in different elevations of the district (Table 17.4).

Table 17.4: Distribution of Area not Available for Cultivation (Percentages)

Size Classes of Population	*Name of Area under Police Station*			
	Garubathan	*Kurseong*	*Siliguri*	*Kharibari*
Less than 200 very small	–	9.29	17.24	15.61
200–499 small	12.28	1.26	22.88	16.33
500–999 low medium	45.26	10.92	32.12	18.11
1000–1999 medium	22.78	17.83	37.82	19.49
More than 2000 large	37.41	17.36	32.06	9.03

The percentages of area not available for cultivation are 29.03 per cent, 18.31 per cent, 24.68 per cent and 17.31 per cent in areas under Garubathan, Kurseong, Siliguri and Kharibari P.S. of the district.

The percentage of area not available for cultivation is on the low side in case of small settlements, which increases towards the larger settlements with minor changes, in different elevations. The medium and large settlements have a large number of infrastructures facilities, and different kinds of welfare arrangements. These settlements have also extensive residential areas. That is why these *mouzas* have high proportion of area not available for cultivation. Relatively small settlements have small number of infrastructures, facilities, etc., so that these settlements have small proportion of that area. Keeping parity with proportion of area not available for cultivation, the density of population is also directly proportional to the settlements with different size classes of population. The large settlements have large number of infrastructures, facilities, extensive residential areas and high density of population and vice-versa (Table 17.5).

Table 17.5: Density of Population per km²

Size Classes of Population	*Name of Area under Police Station*			
	Garubathan	*Kurseong*	*Siliguri*	*Kharibari*
Less than 200 very small	5	33	66	102
200–499	30	34	222	185
500–999	61	292	441	294
1000–1999	112	194	175	531
2000+	248	574	510	1008

The proportions of net sown area are 63.36 per cent, 50.27 per cent, 65.15 per cent and 57.16 per cent in areas under Garubathan (3000 m–5000 m), Kurseong (1000 m–4000 m), Siliguri (500 m–750 m) and Kharibari P.S. (250 m) respectively. The proportion of net sown area is thus inversely correlated with the elevation. In other words, higher the elevations lower the proportion of net sown area and vice-versa.

Table 17.6: Distribution of Net Sown Area

Size Classes of Population	*Name of Area under Police Station*			
	Garubathan	*Kurseong*	*Siliguri*	*Kharibari*
Less than 200 very small	–	58.92	74.19	63.61
200–499 small	57.37	90.49	67.86	60.23
500–999 low medium	46.96	50.71	56.02	55.91
1000–1999 medium	70.52	46.86	53.86	52.35
2000+ large	56.39	52.27	55.31	61.93

On the plain part of the district (Siliguri, Kharibari) the size classes of population of settlements and proportion of net sown area are inversely proportional to each other, indicating large settlements have smaller proportion of net sown area and vice-versa.

That means in plain terrain all settlements have large proportion of net sown area, because these settlements have small proportion of wasteland and area not available for cultivation. On the other hand, the large settlements have high proportion of waste land and area not available for cultivation, so the proportion of net sown area is small.

On the elevated part of the district, (Gatubathan, Kurseong) almost same trend is diacernible, although it is not so distinct on the plain terrain. As for example the proportion is nil in very small settlements of areas under Garubathan P.S. where as medium sized settlements have 70.52 per cent of their land as net sown area, although large *mouzas* have 56.39 per cent of land as NSA. So, the above mentioned trend is not very distinct in case of area under Garubathan P.S.

Irrigation is not available in the southern plains of Darjeeling district, but it is available in the northern mountainous part. In areas under Garubathan P.S. (3000–5000 m) no relationship can be established between proportion of net sown area and proportion of net irrigated area in settlements with different size classes of population, as for example, small settlements have high moderate proportion of NSA *i.e.*, 57.37 per cent which is supported by high proportion of net irrigated area (48.90 per cent), but the medium settlements have as high as 70.52 per cent of the land as NSA, whereas percentage of net irrigated area is as low as 21.83 per cent. Moreover, the large settlements have 56.39 per cent of their land as NSA out of which only 13.28 per cent receive irrigation. However, proportion of net irrigated area and settlements with different size classes of population are inversely proportional to each other on the northern elevated part (Garubathan, Kursaong) of the district.

Unlike in areas under Garubathan P.S., in area under Kurseong P.S. the percentage of NSA and percentage of net irrigated area are more or less directly proportional to each other, and both these variables are inversely proportional to the settlements, with different size classes of population. That means in smaller settlements both the proportions of net sown area and net irrigated area are on the high side which decrease towards the larger settlements (Tables 17.6 and 17.7).

Table 17.7: Distribution of Net Irrigated Area

Size Classes of Population	*Name of Area under Police Station*			
	Garubathan	*Kurseong*	*Siliguri*	*Kharibari*
Lass than 200 (very small)	–	68.04	–	–
200–499 small	48.90	64.79	–	–
500–999 low medium	–	49.01	–	–
1000–1999 medium	21.83	27.64	–	–
2000+ large	13.28	40.52	–	–

Conclusion

The above analysis draws following conclusions:

The forest resources are concentrated in the north-western, part of the district which are highly elevated, rugged and covered by hill soils. The cultivable wastelands are concentrated in the southern part of Darjeeling district where the elevation is moderate to low and covered by hill and new alluvium soils. Areas not available for cultivation are mainly found in large quantities in the north eastern, eastern and south eastern parts of the district, where high concentration of commercial activities are observed. Net sown area is largely found in the southern part where vast stretches of plain land is available, and are covered by new alluvium soil, but irrigation facilities are rare, there.

The area with high concentration of forest resources are inversely related with areas not available for cultivation, cultivable wasteland and net sown area, but the same is directly proportional to net irrigated areas. Areas not available for cultivation are positively correlated with net sown area, cultivable wasteland and net irrigated area, but it is negatively correlated with forest lands. Net sown area *i.e.,* directly proportional to area not available for cultivation and cultivable wasteland, but inversely proportional to forest lands and net irrigated areas. Cultivable wastelands are positively correlated with area not available for cultivation and net sown area, but are negatively correlated with forest lands and net irrigated areas.

In the most elevated part very small settlements have high concentration of forests, but large settlements have concentration of the same in moderately, elevated part of the district. Forests are not found in plains of the district. Cultivable wastelands are inversely proportional to settlements with different size classes of population on the elevated part, but are directly proportional with the settlements on the plains. Area not available for cultivation is directly proportional to settlements with different size classes of population in both the elevated and plain parts of the district. Density of population also shows same relationship. Net sown areas are inversely proportional to the settlements with different size classes of population in both the mountainous and plain topography of the district. Net irrigated areas show same relationship on the mountains, but it is rare in the plains.

References

Directorate of Census Operation, 1971. *District Census Handbook*, Darjeeling District, Part X(A and B).

Lahiri, Sulekha, 1981. Landuse in Eastern Himalayan region: A case study. In: *Perspectives in Agricultural Geography*, (Ed.) Noor Mohammad. Concept Publishing Company, New Delhi, pp. 305–329.

Mandal, R.B., 1982. *Land Utilisation: Theory and Practice*. Concept Publishing Company, New Delhi, p. 129, p. 244.

Appendix–I
Land Utilisation in Darjeeling District

Sl.No.	Name of Police Station	Per cent Forest Land	Per cent Cultivable Wasteland	Per cent Area Not Available for Cultivation	Per cent Net Sown Areas	Per cent Net Irrigated Area
1.	Sukhiapokhri	68.47	2.24	5.30	23.98	8.76
2.	Pulbazar	41.25	5.09	5.71	47.93	11.42
3.	Darjeeling	31.47	3.14	10.88	54.49	19.99
4.	Rangli Rangliot	36.11	3.43	2.66	57.78	6.00
5.	Jore Bunglow	37.76	9.01	11.25	41.96	7.77
6.	Kalimpong	4.16	13.66	23.24	58.92	22.17
7.	Garubathan	0.95	6.63	29.03	63.36	19.30
8.	Kurseong	15.52	15.88	18.31	50.27	38.63
9.	Mirik	14.67	15.57	12.02	57.71	6.04
10.	Naxalbari	–	1.59	16.69	81.70	–
11.	Siliguri	–	10.16	24.62	65.15	–
12.	Phansidewa	–	23.43	12.69	63.86	–
13.	Kharibari	–	25.51	17.31	57.16	–
14.	Darjeeling district	20.45	10.42	15.09	54.04	12.99

Chapter 18

Environmental Impact of Unhindered Land Use in a Village in Darjeeling Himalaya

Jyotirmoy Sen and N.D. Lama

ABSTRACT

Alubari Basti, which fringes the eastern periphery of Darjeeling Municipal Area, was surveyed by the authors essentially to study its multi functional character. A house-to-house survey, incorporating the questionnaire method was adopted to have an in-depth geographical understanding of the study area. I sequential and comparative mapping of land use was done for 1986, 1996 and 2001: using the mouza map as the base map. Significant land use changes that have taken place, primarily due to the pressure of excessively increasing population have been noted.

Introduction

Darjeeling, the northernmost district in West Bengal, lies between 26° 31′N and 27° 33′N latitudes and between 87° 53′N E to 87° 59′E longitudes. Darjeeling is derived from the word "DORJEELING" which means "the land of the thunderbolt".

Darjeeling district is bounded on the north by the state of Sikkim, on the West and north east by the international boundaries of Nepal and Bhutan respectively; while the eastern section is flanked by the district of Jalpaiguri. The southern section neighbours Bihar and North Dinajpur.

Alubari Basti lies in the eastern periphery of the municipal boundary of the Darjeeling town. The main reason for the selection of this particular site was prompted by the multi-functional nature of the area. A total of 161 households were surveyed, from which the following analysis was made:

Households Break up of Male-Female Population

In India males out number females. The same is true within West Bengal and Darjeeling District. A similar trend was noted in Alubari Basti. In the 161 households surveyed, the total population was 756 persons, of which 380 were males and 376 females. In most households, the head of the family was the father, except in 16 households, where the widowed mothers headed the families.

Age-Sex Ratio

The Age-Sex pyramid exhibits the typical triangular pyramid with a larger population in the younger age group and a decrease with increasing age. The age group 10-49 years has the highest population.

Religion

Of the 161 households surveyed in Alubari Basti 121 households were Hindus, 35 Buddhists, 3 Muslims, 1 Sikh and 1 family was Bahai. Some families had backgrounds of mixed marriages where the mother was a Christian.

Occupation

The town of Darjeeling is basically multi-functional. In Alubari Basti, 20 households were engaged in Govt. Services with the PWD, Court, Municipality, Police Deptt. along with various banks. In 20 households, the heads of the families were teachers. In some families, the ladies were teachers. The non-working dependent populations comprised of infants, children, full-time students, illiterate persons, disabled persons or unemployed youth seeking employment for the first time or those presently out of work.

Education

The pioneers who spread education on in Darjeeling town were the Christian missionaries. An encouraging trend noted in the course of the survey was that all young children of the area were attending school. Another interesting point was that a number of students had opted for technical courses like engineering, medical as well as vocational courses like polytechnic, tourism, nursing and hotel management rather than pursuing traditional streams. The main reasons for school dropouts were some form of employment or a family business for the boys; while for girls, it was marriage.

Income/Expenditure

Regarding the income structure, we found that most families belong to the middle-income group, where the head of the family was engaged in Govt. service. In most households, income was more than expenditure, but only in families with high incomes could certain amount of money be saved every month.

House Types

In analyzing different house types the following points were noted:

1. The building materials,
2. The size of the houses and
3. The nature of the house.

Wood and tin were the most widely used materials followed by cement and wood or only cement. Most families lived in one or two roomed houses. A few families had more than 6 rooms in their own houses. It was seen that families residing in govt. quarters numbered the largest. Very few families owned houses and the remaining families lived in rented houses.

Sanitation/Water Supply

Regarding sanitation, it was seen that most houses had their own arrangements on a private or communal basis but some families still had to depend on municipal toilets.

Drinking water supply for most of the families was provided by the municipal corporation but many families have to depend on nearby jhoras.

Land Use

The Mouza map of Alubari Basti, reveals an area, almost oval in shape with pointed extensions in the northern and southern extensions. The Rungdung Khola flanks the eastern side, while the Tenjing Norgay Road and a small jhora flank the western side, while Pandam lies to the east of Alubari Basti. A comparison of the mouza in 1986, 1996 and 2001 reveals some very interesting aspects. The most significant land use change is the extensive spread of settlement areas that seem to enlarge from two small pockets in the central eastern sections and converge together, devouring large areas of tea garden, agricultural and forested areas.

In order to meet the increasing demands of these expanding settlements, the agricultural lands have also extended further, wiping out two small patches of virgin forests.

The fallow lands and tea estates are also pushed back marginally. In 2001 a small patch of fallow hnd developed in the southern extremity, as a result of a recent land side.

Conclusion

From the study it can be seen, that even in comparatively sparsely populated regions like Alubari Basti, the rising population is exerting a demand for the expansion of settlements and agricultural lands, at the cost of the natural floral habitats and pristine environment.

It can be noted that the increasing pressure of population on the fragile region will not only jeopardize life support systems for the hill people, but its effects will also cascade down to the heavily populated plains. This calls for a "Watershed Development Approach" in the Darjeeling Hills, which will mould, reshape and stabilize an already fractured environment, working towards an optimum production with minimum pressure on natural resources. This is very important for preserving extremely fragile ecosystems in the Himalayas.

Chapter 19

"Livelihood-Economic-Environment" Approach for Sustainable Development of Himalayan Mountain Regions

Raj Ganguly

ABSTRACT

The sustainable economic development process in the Himalayan mountain region leaves challenges of protecting environmental degradation and long term economic vulnerability of the people of the region. The economic growth, social development and environment conservation are the three pillars of sustainable development, which if not properly integrated, often may mislead the direction of development process.

One common factor for all the three facets of sustainable development is 'people', which is not only at the receiving end of the development process but also plays the dynamic role as a change agent. The 'livelihood' of people is the main driver of development and needs to be sustainable too. Diversifying the livelihood opportunities, beyond agriculture is the much-needed strategy for sustaining economic development. The integrated approach to strike the balance between sustainable economy and sustainable environment, by way of creating and sustaining sustainable livelihoods shall secure a bright sustainable mountain society in a perfect unison with human spirit.

Introduction

Mountains cover 24 per cent of the Earth's land surface, host 12 per cent of its people and supply fresh water to almost half of the world's population (Price, 2004). As sources of water, energy, and agriculture and forest products, and as centre of biological and cultural diversity, mountains are important for at least half of the humanity. Despite this resource richness, mountains have become economically marginal lands where poverty is pervasive. The ruthless exploitation of natural resources in the name of economic development, often from downstream hinterlands have milked the mountain regions to such an extent that now these areas are environmentally vulnerable and pose potential danger zones in the wake of climate change.

Paleologic records indicate that climate warming in the past has caused vegetation zones to shift to higher elevations, resulting in the loss of some species and ecosystems. Simulated scenarios for temperate-climate mountain sites suggest that continued warming could have similar consequences. Species and ecosystems with limited climatic ranges could disappear and, in most mountain regions, the extent and volume of glaciers and the extent of permafrost and seasonal snow cover will be reduced. Along with possible changes in precipitation this would affect soil stability and socio-economic activities such as agriculture, tourism, hydropower and logging. Resources for indigenous populations and recreational activities would also be disrupted (UNDP).

Thus, the climate changes along with the increasing population pressure shall continue to have wide-ranging impacts on mountain ecology, mountain people and on their economy apart from creating serious challenges to the economies of the plain areas. There is increased global attention, in last two decades to the climate change and its' repercussions on mountain ecosystem, suggesting need for searching sustainable development policies for the people, of the people and by the people of the mountain regions.

The present situation of Himalayan ecosystem is very critical. Unchecked deforestation, faulty land use practices, settlement of burgeoning population on mountain slopes, unregulated construction of roads and encroachments of forest lands for the creation of arable lands have caused adverse geo-ecological impact (Ives, 1989). Often the problems and reasons thereof are intertwined, thus making the whole development issue very complex in

Himalayan mountains. The socio-political leadership continues to steer the economic development for the people in the short term but leaves greater challenges of environmental degradation and long term economic vulnerability, for the region and other regions dependent on it.

The 'Gross Domestic Product' and 'Per Capita Income' focused development plans have led to large scale industrialization of Himalayan region. Tax benefits, cheaper lands and investment friendly measures have been the main attractants as the type of industry being established are miles away from the core competence of the region. The cities like Parwanoo, Haridwar, Rudrapur, Jammu etc. has become home for manufacturing chemicals, cars and consumer articles, which seldom enjoys any geographic advantage of resource nearness, but continues to add to the environmental and socio-economic problems. Even the so-called industrialization in the mountain region has failed to create livelihood for mountain people as the very nature of these industries is different from the skill set of the hilly local people. Instead, it has caused in migration from plains adding more social challenges. The multipurpose dams created in the Himalayan regions, to protect the water and energy security for the people of plains, had huge environmental and socio-economic impact in the region. The benefits accrued to local mountain economy, from such large projects are yet to crystallize.

Economic development is as important as the goal towards environment conservation for long term sustenance of humanity. Human beings use natural resources available on earth such as land, water, vegetation, minerals etc. to make a living. As human wants are endless, the economic development keeps on marginalizing natural resource utilization beyond its' replenishing capacity. Judicious management of natural resources is vital in today's world where overexploitation by growing population is happening at a fast pace, pressing a need for sustainability of the development process. Mountains are highly vulnerable to human development and natural ecological imbalance thus calling an urgent attention for holistic sustainable development.

The recent framework for sustainable development is 'that which meets the needs of the present without compromising the ability of the future generations to meet their own needs." (World Commission on Environment and Development, 1987) The three pillars of sustainable development *viz.*, economic growth, social development and environment conservation, if not fully integrated, often mislead the direction of development process. One factor that is common for all these three facets of sustainable development is 'people'. People are the key element, which is not only at the receiving end of the development process but also plays the dynamic role as a change agent. The 'livelihood' of people is the main driver of development. It is the most visible element, which often acts as an incentive to propel the development process in the set direction. The economy influences the livelihood in both positive and negative way, or *vice versa* so far as its' impact on environment is concerned. The economic development ushered in the mountain region although created many new income opportunities, have moved the people away from their traditional livelihoods, thus creating far reaching socio-economic and environmental consequences.

Sustainable Livelihoods

The term sustainable livelihood was used as a development concept in the early 1990's. Chambers and Conway (1991), defined it, as "A livelihood comprises people, their capabilities and their means of living, including food, income and assets. A livelihood is environmentally sustainable when it maintains or enhances the local and global beneficial effects on other livelihoods. A livelihood is

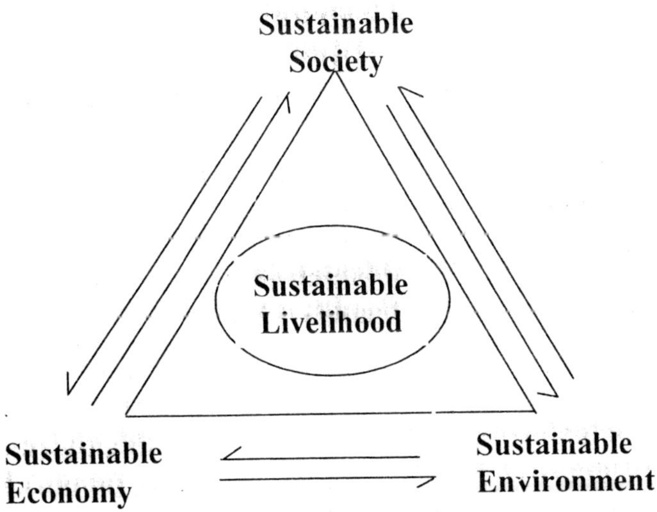

socially sustainable which can cope with and recover from stress and shocks, and provide for future generations."(Chambers and Conway, 1991) For sustainable development, the livelihood needs to be sustainable too. However, the perspective of sustainable livelihood should be more holistic and should denote the livelihood that is in complete harmony with the environment.

The mountain ecology with its' blessed biological diversities had generated typical livelihood patterns which were interdependent on each other and were in complete harmony with ecology. The economic development that underestimated these critical linkages, created huge surpluses on one hand to be marketed in the plains but on the other resulted in ecological imbalance and livelihood insecurity thus causing migration and poverty. The population pressure and climate change further compounded the problem. Large scale migration from mountain areas to plains is the immediate result of un-economical status of livelihood or lesser livelihood opportunities.

Agriculture and livestock has remained the most important livelihood for mountain people. Over the years, the traditional knowledge accrued by the peasants had tuned crop cultivation and livestock rearing in perfect balance with natural ecosystem. The onslaught of natural resources propelled by market demands created misbalance between ecology and livelihood, thus making agriculture un-economical. Large-scale deforestation is mainly responsible for increased landslides and decreased water in the streams. It disturbed the flora and fauna thus impacting the agriculture by increased animal and bird damage to fruit orchards and increased pest problems, fodder scarcity, soil erosion etc.

'Poly-cropping'*, 'mixed farming' systems etc. are time tested agricultural practices used by the farmers in Himalayan Mountains. This provides not only nutritionally balanced food basket and enrich soil but also act as a risk-mitigating factor in changing climatic situations. 'Productivity centric' agriculture have favoured high input requiring varieties. Mono-cropping and cash crop cultivation *viz.*, tea and apples have made the agriculture unsustainable. The Apple orchards of Jammu and Kashmir and Himachal Pradesh,

---

* *Poly-cropping*: A cropping system in which a number of crop seeds (sometime as high as 9 crops) are mixed and broadcasted in the field.

favoured increased input usage thus slowly making the cost of cultivation tread on very thin line of profit, and having huge ecological impact. Over the years, it is evident what socio-economic and ecological benefits the tea estates in the North East have brought to local people and region, except for accruing wealth elsewhere.

Mountain livestock also plays a key economic role providing food, fibre, transportation, and supporting agriculture. However the policy focus on only 'milk' and that too fixing quality parameter as 'fat percentage', created huge impacts on the mountain livestock. Promotion of cross breed dairy cattle, buffaloes across the Himalayan region (http://www.fao.org/docrep/x5862e/x5862e07.htm) changed the livestock diversity to such an extent that fodder problem further worsened and so as quality of manure and draft animals, besides genetic erosion local breeds. There is much promise for strategizing the mountain livestock development towards fibre, meat and draft purposes besides creating market oriented value added product thereof.

Organic farming system, which has been the default practice of mountain farmers since ages, can prove vital to revive mountain agriculture economically. The market for organic products is growing worldwide and the market oriented 'organic product value chain', should bring this important livelihood back on sustainability track. An integrated approach of livestock and agriculture along with mixed farming practices with fruit, vegetable, medicinal plants, fodder etc., will ensure ecological and economical outcomes to create a range of sustainable livelihood practices for the farmers of this region.

Diversifying the Livelihoods

Failure in the sustainability of livelihoods and environmental degradation are intricately connected because the shortage of livelihood options can lead to the destruction of available natural resources (Rawat, 1995). Apart from agriculture there need to be a diversity of livelihood opportunities for the people for sustaining economic development. Economic policies which were designed or promoted without keeping the principal characteristics of mountains, its' ecology, people and resources although created many livelihood opportunities but eventually failed to sustain itself and ameliorate the growing ecological impacts.

The dimension of sustainable development for mountain ecosystem for mountain ecosystem should invariably include Food and feed production system, Health system, Energy, Transport, Waste management, Education and skills, Trade, Business and Industry. Across all the sectors the principles of sustainability should help realize the balance between 'people', 'profit' and 'natural resources'. Sustainable development efforts in line with sustainable economy targeting generation of diversified sustainable livelihoods shall create visible incentives to reach the overall objectives to secure sustainable environment.

The agro-processing industry is the backbone of agrarian economy and as a complimentary industry for organic farming system, organic food processing industry will have a sustainable future in the mountain region. Decentralised development of small food processing facilities across the region shall go a long way in developing a vibrant and sustainable economy for the region, creating diversified and women centric livelihoods. Ayurvedic medicine, herbal cosmetic can be other offshoot industries.

Energy is one important area of sustainable development, if not planned well it can have serious impacts on ecology and livelihood. On one hand, multipurpose dams are designed to cater to the energy requirement of plains largely ignoring the environmental impacts. The mountain peoples' energy requirement still largely depends on biomass. Mountain regions are blessed with water resources and as such, series of pico and micro hydro power units can be planned to not only meet the energy requirement of communities but also to support the small agro-processing units. An integrated approach to mix bio-mass energy through improved tools *e.g.*, bio-gas, smokeless ovens etc., renewable energy sources like solar, wind and water can meet the energy requirement of mountain people. It should well be complimented with energy efficient 'building', 'processing system' and 'appliances' etc.

Roads and transportation are the vehicles of economic development, apart from those for security reasons. It is equally accountable for the ecological degradation in the mountain region. The landslides, soil erosion and destruction to the flora and fauna of mountain region all is attributed to roads. Attention needs to be given to the research and development in this area such as road building material suitable in mountain terrain, alternative transport like gravity cable cars etc. The minimum emission levels of transport vehicles, mandatory fuel norms eg., CNG, low sulphur diesel, should take policy precedence in mountain regions.

An industry like Tourism etc. also generates livelihoods but is not sustainable, unless planned differently for mountains. In most cases, it has created 'in-migration' of non-mountain people with less ecological sensitivity, thus further adding the social and environmental challenges. Eco-tourism, health tourism etc. can go a long way to make the mountain tourism sector much more economically stronger, sustainable and eco friendly. The handicrafts industry developed parallel to tourism industry can have synergistic effect on sustainable development.

The health system of the mountain region with traditional health system nearing extinction and the modern system failing to reach the remote corners is in critical situation. It requires a paradigm shift in approach. The health infrastructure as envisaged in other areas may not fit well in the context of mountain regions. Instead, 'Mobile Health Clinics' can be planned as a support infrastructure to 'Primary health centres' at block level, to take the health services to remote areas. Efforts towards strengthening and reviving the traditional medical system should help create livelihood apart from finding new opportunities such as 'health tourism'.

Waste management system is another critical challenge for sustainable development. Planned urban and rural development should encompass a suitable self-recyclable system of drainage and waste disposable system. The banning of non-biodegradable material such as 'plastics' is a must in the mountain region to stop further ecological hazard. It is high time to universalize the waste disposable norms in the mountain region, benchmarking against the international best practices.

Business and trade in the mountain regions for various mountain products and natural resources are important for economic development and prosperity of local people. However, these should be within the purview of sustainable economy principles so as not create stress on the ecology. The sustainable business and trade should beget sustainable livelihoods, rather than only churning wealth. It should stand on the principle of resource optimization, production optimization and profit optimization within the framework of sustainable development perspectives. The integrated approach of sustainable development by keeping the balance between sustainable economy and sustainable environment, by way of creating and sustaining sustainable livelihoods shall secure a bright future for mountain regions, its' ecology and people in particular and for the humanity in general. It shall go on to create a sustainable mountain society in perfect unison with human spirit.

References

Chambers and Conway, 1991. http://www.livelihoods.org/SLdefn.html.

http://www.fao.org/docrep/x5862e/x5862e07.htm. Livestock in mixed farming systems of Hindu Kush Himalayas, FAO Corporate Document Repository, Forestry Department.

Ives, J.D. and Messerli, 1989. *The Himalayan Dilemma; Reconciling Development and Conservation*. Routledge, The United Nations University.

Rawat, A.S., 1995. Deforestation and forest policy in the Lesser Himalaya Kumaun: Impact on peasant women and tribal populations. *Mountain Research and Development*, 15(4): 311–322.

Price, Martin F., (Ed.), 2004. *Conservation and Sustainable Development in Mountain Areas*. IUCN, Gland, Switzerland, and Cambridge, UK, 29 pp.

UNDP/GRID. Arendal; http://www.grida.no/climate/vital/28.htm.

Chapter 20

Global Warming and Melting of Himalayan Glaciers

R.D. Gupta

ABSTRACT

Increasing concentrations of greenhouse gases like carbon dioxide (CO_2), methane (CH_4), oxides of nitrogen (Nitric oxide: NO; Nitrogen dioxide: NO_2; Dinitrogen oxide: N_2O; Nitrogen: N_2), chlorofluorocarbons (CFCs), ozone (O_3) etc. in the atmosphere warrant *"global warming"*. Their higher building-up in the atmosphere is likely to lead to temperature rise of 1.1 to 6.4° Celsius and change in precipitation pattern over the next 50–75 years or by the end of twenty first century. *Global warming* will not only spawn flooding, drought and drastic reduction in crop production but also results in the melting of snow from mountain tops and glaciers. The gradual rise in temperature all over the world has resulted in the melting of glaciers in the Arctic and Antarctic oceans as well as that of glaciers existing in the Himalayas. In this chapter of the book, however, melting of Himalayan glaciers has been briefly elucidated.

The study reveals that majority of the Scientists–Environmentalists/Geologists and Glaciologists–are of the firm view that glaciers are receding, which are, thereby, threatening the water supplies in the rivers like the Ganga, the Brahmaputra and the Indus. Surface area of many of the Himalayan glaciers has shrunk by 21 per cent since 1962. World Wide Fund for Nature has predicted that the glaciers are melting due to *global warming*. Satellite imagery results have shown that average recession of the Gangotri glacier is about 25 m year^{-1} for the last three decades. Accelerated melting of this glacier will lead to flooding, landslides and reduction in river's flow. Production in agriculture and hydropower generation would be badly affected. Similar is the situation of Siachen glacier which will eventually hit the flow of Indus river's water. Fate of glaciers of Kashmir Himalayas which feed the Jhelum, the Chenab, the Lidder and many others, has also been described. Remote Sensing Agencies have shown an increase in the formation of glacial lakes in Himalayan region, especially in Himachal Pradesh, which spread in the basins of the rivers--Sutlej, Ravi, Beas and Chenab. In the end, suggestions and priorities for the future have been presented.

Introduction

It is an established fact that the planet Earth receives energy from the sun in the form of radiation. About 30 per cent of the incoming radiation reflects and the remaining 70 per cent is absorbed by the land, oceans and atmosphere to warm them. Absorption of this radiation helps the biodiversity (various life forms) to exist and flourish on the Earth. But there are certain gases like CO_2, CH_4, CFCs, O_3 and NO which absorb infrared radiation going out from the Earth, thereby increasing atmospheric temperatures across the world. This is known as "Greenhouse Effect". The word "Greenhouse Effect" has been derived from greenhouse gardens which trap heat and help some plants to grow in the cold or cold arid regions of the world.

Thus, an increase in the average temperature of Earth's surface air and oceans due to the "Greenhouse Effect" is known as *"global warming"*. *Global warming* may trigger a number of events that could lead to global changes in the climate. It has been estimated that an average global surface temperature which is presently about 15 to 16° Celsius would increase by 2 to 3 or 3 to 4° Celsius or by 1.1 to 6.4° Celsius (Shiva, 2007) during the end of 21st century. The effect of this phenomenon could even be seen more in the polar regions (North Poles and South Poles). The consequences of *"global warming"* would be dire, ranging from sea level rise to melting of glaciers. Increased intensity of weather phenomenon like cyclones, changes in amount and pattern of precipitation, alterations in the yield of various foodgrain crops, fruits and vegetables, increase in the ranges of insect/disease vectors to extinction of various species are the other effects of *global warming*. The chapter however, deals with an impact of "*global warming*" on the melting of "Himalayan Glaciers".

Melting of Himalayan Glaciers

Global warming is set to render a terrible mark on the high altitudes. Whereas a change in climate is taking heavy toll on the giant glaciers of the Himalayas by way of their melting, they are giving menace to the lives of thousands of native people and host of animal species (Sharma, 2000). *Global warming*, infact, would impact 5018 glaciers in the Himalayas.

Glaciers in the Himalayas, which feed important rivers such as the Ganga, the Indus, the Brahmaputra, responsible to provide water to millions of people for drinking purposes as well as for irrigating the crops and industrial uses are unfortunately shrinking due to *global warming*.

In the Himalayas, which hold the largest glaciers or the highest ice reserves after the polar ice caps, the rate of their melting is the fastest. With thousands and lakhs of square kilometer of Himalayan glaciers, these are aptly called the water tower of Asia and ensure round-the-year supply of water to billions of people. Recent research, however, shows that 67 per cent of the Himalayan glaciers are retreating at a startling rate. But, the shrinking of the Himalayan glaciers has become much faster during the last three decades than they did during the last 200 years (Sinha, 2005). The melting glaciers form natural lakes whose sudden rupturing is a cause of great concern. As the ice melts, lakes are developed between the mounds of debris and rocks that are left behind by the sliding glaciers.

According to many scientists, the worldwide glacier melting rate has doubled (Khan, 2007) since 1980. It is not surprising with the fact that the 1990s was the hottest decade of the Earth during the last 1000 years. The study carried out in the ten sub-basins of Indus River System, namely Swat, Chetral, Gilgit, Hunza, Shigar, Shyok, Upper Indus, Shingo Aster and Jhelum identifies about 5218 glaciers covering an area nearly 15040 km^2 (about 11.7 per cent of the total geographical area often basins).

Extent of Himalayan Glaciers and their Influence on Weather and Climate

Extent

It is well known that over the Indian Himalayan Region about 5000 glaciers exist, which sprawl over an area of 43,000 km^2. Thus, an average size of glacier could be worked out as 8.6 km^2. However, an actual size may range from low as 0.5 km^2 to as high as 200 km^2 in Kashmir known as Siachen (Upadhyay, 1999).

Influence of Glaciers on Weather and Climate

Snow covers and glaciers have their own effect on weather and climate which is described briefly hereunder:

1. Snow and glaciers modify the local weather/climate. The short-term effects are the formation of cold air mass at lower levels. Due to high albedo, high emissivity, poor conductivity and low melting point, snow acts as a strong cooling agent. Due to cooling effects excessive snow accumulation over Western Himalayas may delay development of seasonal summer low over Northwest India, which is the driving force for Southwest monsoon.
2. Glaciers are indicators of climate and climatic change. A glacier mainly composes of stratified layers of successive deposition of snow cover over a long period of time. Inside its numerous strata, sediments and life forms–seeds, bacteria, pollen etc. are buried.
3. Ice core data are also useful for monitoring the level of pollutants. Variation of CO_2 level seems to be closely associated with the glacial oscillations.
4. Avalanche is a natural disaster of snow-covered mountains generated by structural failure of snow-cover resting on steep slopes.
5. Glaciers feed all the perennial rivers which originate from the Himalayas. In India, the contribution of the Himalayas to the Indian River System is to the extent of 20 per cent. Snow storage can be used for summer water supply for domestic purpose, agricultural requirement and also for generation of hydro-electric power.

General View of Himalayan Glaciers

Glaciers are an object of awe and very important from an ecological and climatic perspective. But now it is well known that the glaciers of Himalayas are considerably shrinking owing to *global warming*. The snout of the Siachen glacier has shown perceptible changes. Similarly from the last thirteen years, the glacier of Gangotri, one of the largest in the Himalayas, has been retreating rapidly.

Recently, scientists of the Space Application Centre in Ahmedabad studied and examined the data for 466 glaciers of Himalayas and found that their surface area has shrunk by about 21 per cent since 1962. As the glaciers retreated, they have become more fragmented, so the fragmented or the smaller glaciers were more sensitive to *global warming*.

In the Tibetan Plateau (the source of major Asian rivers like the Brahmaputra and the Indus), the greatest retreat of glaciers has occurred since the mid-1980s (Sinha, 2005). Between 1950 and 1970, 53 per cent of the glaciers here were receding. While in 1990, 95 per cent were found to be receded.

As the Himalayan glaciers are facing with the danger of receding and causing havoc with the regional ecology, so India has taken-up the matter with China for a joint solution to the problem of Climatic Changes (Anonymous, 2007). The melting of Himalayan glaciers is an important issue because it can have catastrophic effect on the ecology of not only of Indian region but whole of the world. Scientists of Chinese Academy of Sciences have found rapid melting of glaciers of the Tibetan Plateau. The melting of the glaciers will ultimately trigger more droughts, more expanse of deserts and increasing of sandstorms.

According to Working Group II of the Intergovernmental Panel on Climate Change, the Himalayan glaciers are receding faster than any part of the world. The geologists of Geological Survey of India after carrying out an exhaustive study have come to the conclusion that the Himalayan glaciers are receding on an average by 15 m year^{-1}. Actually, hundreds of Himalayan glaciers are shriveling up in varying degrees (Vasudeva, 2007). A receding rate of some of the glaciers has been mentioned in Table 20.1. The data indicates that the rate of receding of glaciers is the maximum in case of Bera Shingri and Meola glaciers followed by that of Zemu and Gangotri (Table 20.1).

Table 20.1: Receding Rate of Some of the Himalayan Glaciers (Average Values)

Name of the Glacier	*Receding Rate (m year^{-1})*
Gangotri	25
Pindari	23
Dokariani	18
Bera Shingri	36
Meola	36
Sona Pani	17
Milam	13
Zemu	28

Source: Vasudeva, 2007.

In the next couple of decades, the Himalayan glaciers may shrink to about a fifth of present size or area and many others including the ones feeding the river Ganga may disappear if the current pace of *global warming* persists. The latest studies done by glaciologists at the Wadia Institute of Himalayan Geology have pointed out a far grimmer scenario. According to Dr. J.T. Gergan, a Senior Glaciologist, who has been studying the Dokriani glacier on the Bhagirathi basin since 1997, there will never be any massive flooding from melting glaciers in the Himalayas. We are already in the period of reduced run-off.

Now many glaciers have come down almost at 2100 m in the Himalayan region. Approximately 60 to 70 per cent of the more than 200 glaciers in the Bhagirathi basin may be reduced to small patches of ice in less than 50 years from now. Only the large ones at higher altitudes like Gangotri might survive. The Yamuna basin which is fed by 52 glaciers is already showing signs of distress. This is the reason why even in a year of good snow, there is shortage of water in the river Yamuna, thereby, causing scarcity of drinking water in Delhi. Just, there is not enough water in the river during the months of March to June, prior to starting monsoon.

World Wide Fund (WWF) for Nature has predicted that the glaciers are melting due to *global warming* and this has been further supported by United Nations Environment Programme (UNEP), based on field studies carried out in the glacial regions. According to WWF, after melting glaciers, first there will be floods on large-scale, and thereafter, an acute shortage of water would occur. Now the question arises, where is water of glaciers going? The glacier erosion carries with it different types of Morains. The last one is called Terminal Morain, after that there is no glacier and only water. As already stated the glaciers have come down at about 2100 m in the Himalayan region. Obviously the melting of ice forms water and small streams are originated to become a part of a big river, for example, the river Aru is a part of the river Lidder in Pahalgam which in turn is a part of the river Jhelum. Similarly, so many small rivers join to make the river Chenab and the Sindh. Thus, slow and normal melting of glaciers guarantee the perennial flow of water in the Himalayan rivers *e.g.*, the river Ganga from Gangotri.

The melting of Himalayan glaciers due to *global warming* will lead to severe floods and shortage of drinking water in South Asian regions (Bhardwaj, 2005). Cumulatively this melting of glaciers could change the way we know our world. If *global warming* is not arrested, the rivers would be dried-up after flooding level of seas will rise and fertile lands will turn into wastelands (Vasudeva, 2007).

The concern of climate change and *global warming* has been highlighted in *India Today* (November 6, 2006) that the length of Samudra Tapu (one of the largest glaciers in the Chandra Basin of Lahaul and Spiti) has been reduced by 862 m between 1962 and 2006 at the rate of 18.5 m $year^{-1}$. Moreover, 127 small glaciers have shrunk by 38 per cent and large ones by 12 per cent (Sidhu, 2006).

Gangotri Glacier: Some Facts

The well-known Gangotri glacier from where the river Ganga originates, which has been receding 20 m $year^{-1}$ for the past 10 years (Gupta, 2005) is now retreating at either 25 m $year^{-1}$ (Vasudeva, 2007) or at 30 m $year^{-1}$ (Shiva, 2007). The Geological Survey of India has been conducting the studies on the recession of the Gangotri glacier since 1935. Accordingly, the rate of shrinking of 30.2 km long Gangotri glacier in the last three decades has been found to be more than the rate during the early decades. The average rate of recession has been computed and 2001 Satellite Imagery results show that the average recession for the last three decades is about 25 m $year^{-1}$ (Anonymous, 2007).

Although the shrinking of the Gangotri glacier is a matter of great concern yet the prospects of its disappearance does not exist. It was stated by the Minister for Science and Technology Shri Kapil Sibal in the Lok Sabha, New Delhi, India. Sibal, however, expressed optimism over the future of this process and the fate of Gangotri glacier. Earlier during the year 2006, in the Intergovernmental Panel on Climate Change, a gathering of scientists and climate experts from across the globe, had portrayed a negative future for the globe in wake of the *global warming*.

Accelerated Melting of Gangotri's Glacier and its Consequences

A survey report of March, 2005, prepared by World Wide Fund (WWF) for Nature in collaboration with India, China and Nepal indicated that in the river Ganga, the loss of glacier melt water will reduce from July to September flows to two-thirds. This will cause shortage of water for 500 million people and 37 per cent of India's irrigated land.

Accelerated melting of glaciers will rise river's water level over the next few decades initially, leading to higher flooding and landslides. But in the long run as the volume of ice available for melting diminishes, a reduction in glacial run-off and rivers flows can be expected.

There are forecasts that upto a quarter of the global mountain glacier mass could disappear by 2050 and upto half could be lost by 2100. A reduction of water flow in the river Ganga will affect the agricultural production as all of the agriculture in north India is highly vulnerable to any change of river flow will have serious implications for the country's economy. Moreover, the entire system the hydropower generation situated on the river system would also be jeopardized both in Nepal and India.

Siachin Glacier: Its Melting and Effects

Siachen glacier, is the largest "Alpine Glacier" on the Earth and the second largest among the other glaciers in the world. The high altitude area around Siachen is precipitous and glacial with height ranging from 3600 m to 7200 m. It is found in the Karakoram mountain at the western end of the Himalayan mountain chain. This mountainous region has the name Siachen which means garden of roses. It was named because the flowers of roses once used to bloom in the valleys far below the peaks.

The melting waters of the Siachen glacier constitute the source of the river Indus, a critical source of water for both India and Pakistan. But it is unfortunate that Siachen glacier nowadays has receded more than 800 m during the last 18 years *i.e.*, from 1988 to 2006. It is attributed to become the highest battlefield of the world apart from the *global warming*. With more than 3,000 troops living and operating there, hundreds of machines, scores of choppers which daily during 1999 Kargil war, between India and Pakistan (Gupta, 2005), deteriorated whole of its environment. Environmentalists have stated that the military activities have sent tonnes of chemicals on the surface of the glacier, thereby, polluting, the headwaters of the Indus river. However, in-house study done by Army showed that 75 per cent of the waste generated is biodegradable including human waste (Morup, 2005). An Army press statement further says, "Non-toxic part of the non-biodegradable amounts to only 1.3 per cent which is destroyed *in situ*." In order to deal with the human waste each post has toilet with high tech biodigester. The equipments which have been transported to the height of 2300 m, bum and destroy the human waste.

Despite such claims made by the Army to preserve the "snow-white colour of the glacier", while holding this highest battlefield of the world seems to be more than mere words of promise. It is further substantiated by the report of World Wide Fund for Nature prepared during 1997, which has estimated that apart from fast melting of the Siachen glacier more than 30 per cent of the regions edemic flora has become threatened. Some species are believed to have become extinct. Several species such as Tibetan gazelle, are understood to be on the bank of extinction.

Such concerns from Ecologists and Environmentalists have now prompted the Governments of India and Pakistan to reduce their military presence in this region. And jointly start the task of regenerating the biodiversity by setting-up a peace park there.

Fate of Glaciers of Kashmir Himalayas

The hydrological data of all the main rivers of Kashmir Himalayas and their main tributaries have revealed that the discharge of these rivers and tributaries has almost become half these days as compared to only 50 years back. Whether it is the Jhelum, the Chenab or it is the Lidder, Sandren, Vishow, Rambiara, their discharge has diminished. There are number of reasons for this diminution of river discharge. The most important among them is rising of normal temperature and decreasing of rainfall/snowfall during winter. Till recently the snowline was just above Pahalgam and Sonamarg which now has receded back to Sheshnag area around 5,000 m altitude. Not only this most of the glaciers of Great Kashmir Himalayan region, which range from Harmukh to Drungdrung including Thajwas, Kolahoi, Machoie, Kangrez and Shafat, have far receded back during the last 50 years. Same is true on the Chenab side of this Himalayan range as most of the glaciers clearly depict that they have drastically receded during the last few decades.

It is believed that discharge of the river Chenab has now become less than half of what it was 60 years ago and the discharge is diminishing year after year. Although no reliable or sound data for dependable discharge of the Chenab's basin as well as the receding of the snow/glaciers is available yet it has been found that fifty years back the Chenab basin had about 8000 km^2 area under glaciers and permanent and ephemeral snow cover. But these days only 4,100 km^2 area has remained under glaciers/snow-cover. Most of the snow fields and glaciers except Nun Kun have reduced to half or even less which is the main cause of diminution of discharge in the river Chenab.

In Pir Panjal range there is hardly any remnant of glaciers on the top of this mountain range, when the Terminal Morains at Akhal (Rajpora) in the Romshi river, the to Dubjan in the Rambiyara river and Guratan in Vishow river bear the testimony to that the glaciers were once extended up to these places in the recent past. If the situations will continue like this, all rivers flowing from the Pir Panjal range will vanish permanently and become ephemeral. Sonamarg meadow is another Terminal Morain which shows that the Thajwas and Baltal glaciers were once extended upto this spot (Shah, 2007).

Glacial Lakes

As the glaciers retreat, they leave behind debris called Morains at their snout, which act as a dam for the melting snow to collect. This is called Glacial Lake. More than 50 lakes caused by glacier movement have appeared worldwide during the past 150 years (Bahadur, 1999).

Remote Sensing Agencies have noted an increase in the formation of glacial lakes in the Himalayan region. A study conducted by the scientists of Himachal Pradesh Agriculture University in collaboration with those of UNEP and the International Centre for Integrated Mountain Development (ICMOD), Kathmandu, indicated a large number of glacial lakes in Himachal Pradesh. Out of them, 229 were very prominent. Of these glacial lakes, 22 spread in all the major river basins of Sutlej, Ravi, Beas and Chenab, which have been identified as potentially dangerous. There have been a number of Glacial Lake Outburst Flows in the past few decades in the upper altitudes, which affected local communities largely. Some of the recent flash floods in the Sutlej and the Beas could have been due to Glacial Lake Outburst Flow. As these lakes get bigger people in more populated downstream areas will be impacted. The Pareechoo Lake burst led to losses worth Rs. 8000 millions (Rs. 800 crores) as reported by Sinha (2005).

Melting of Glaciers of Antarctica

The Himalayan glaciers should not be viewed singularly in respect of their melting. But there are Alpine glaciers, Alaska glaciers, Rockies and Andes glaciers as well a large snow sheets of Greenland and Sibarian Antarctica and Arctic region, which are also melting. So the effect of *global warming* will not be restricted to Himalayan glaciers only but other glaciers of the world as well.

Two separate studies from climate researchers and the researchers of Space Agency showed that the glaciers are flowing into Antarctica's Weddell Sea (Anonymous, 2004). Their satellite measurements indicated that climate warming can lead to rapid sea level rise. The scientists of National Academy of Sciences America found that parts of Antarctica are steadily melting. Large ice shelves in the Antarctica Peninsula disintegrated in 1995 and 2002 as a result of climate warming. These floating ice shelves did not affect sea level as they melted. A drastic change in just 15 years was seen in 240 km of coastline. The affected area was at the far nothern tip of the Antarctica, just south of Chile and Argentina. Temperatures have risen by 2.5°C in these areas during the last 63 years which are faster than almost any region in the world. In the past 33 years ice shelves have lost more than 13500 km^2 area in this region.

Suggestions and Priorities for the Future

1. Enhancing afforestation will certainly reduce the CO_2 level, thereby, lessening the Greenhouse Effect. As trees are the big natural "sink" of CO_2, they can utilize CO_2 through photosynthetic process.
2. While enhancing afforestation *i.e.*, planting of the trees, the rural forestry and urban forestry programmes must be encouraged. Local species of plants/trees should be preferred to grow.
3. Felling of trees should be totally stopped. Stern action should be taken against the defaulters.
4. Greenery and reforestation of the landscape is the only and immediate solution to absorb the ever increasing concentration of CO_2 and, thereby, decreasing, Greenhouse effect.
5. As the glaciers are receding at such a fast rate that most of the glaciers will disappear within few decades, so suitable measures for their conservation and management should be taken.
6. It is suggested that military movement in Siachen be reduced. It would result in reduction of human suffering and help save money. Reducing military movement will also prevent in disturbance of wildlife that will consequently assist in restoring environment and ecology.
7. A "Siachen Science Centre" should be set-up which can help in advancing, the cause of science in many fields. The Siachen glacier has the potential to be the highest altitude manned research station in the world. Moreover, because of its isolated location, unique geology and geographic position, it would be a special place for carrying out research in many fields like astronomy, geology, atmospheric sciences, glaciology, life sciences, psychology and behavioural sciences.
8. There is need to focus on concrete climate change adaptation measures. This reduces loss of life and habitat, and finally improve the food security and health conditions.

9. Avoid the crowds from visiting the sources of the Ganga, the Yamuna, the Satluj, the Beas, the Indus, the Brahamputra, the Jhelum and the Chenab (The Chandra and the Bhaga), the Krishna, the Godavari, the Kaveri and others. The crowds completely disturb and pollute the environment and ecosystem of the region.
10. The containers particularly the polythene bags and tins of small and medium sizes, kerosene stoves, gas cylinders, refrigerators, are all natural hazards to the environment of glaciers and glaciology. Use of these practices, therefore, should be minimized.
11. In addition, large number of people use temporary toilets, take baths and use soaps and detergents which create pollution in the region and facilitate the melting of glaciers. Used clothes, towels, *dhoties, sarees* and bathing suits and other items of human use are thrown there, thus creating huge hazards for glaciers and create conditions for them to recede. Recently, tonnes of waste material were collected from Amarnath (Jammu and Kashmir) pilgrimage and similar may be the conditions for Kedarnath, Badrinath, Yamnotri and others.
12. It has been found that if the developing countries make more provisions of irrigation facilities and change planting times/days of crops and take other measures to adapt to *global warming*, production of cereals like rice and wheat would drop from 9 to 11 per cent and 5 to 7 per cent, respectively by 2060.
13. Advanced researches are essential to assess and evaluate the potentials of methanol, obtained from methane, which can be effectively used in transport sector.
14. There should be restrictions on the emissions of dangerous CO_2, CH_4, CFC's from the factories and automobiles.
15. Biogas is another alternative source of conventional energy for domestic use. The conversion of cow dung from hearth to the biogas plant will not only supply fuel but will serve as a natural fertilizer to the agricultural plants.

References

Anonymous, 2004. Antarctica glaciers melting faster: Study. *The Daily Excelsior*, 40(266): 7.

Anonymous, 2007. Gangotri glacier receding by 83 feet every year. *The Kashmir Times*, 64(243): 16.

Anonymous, 2007. Melting of glaciers: India takes up matter with China. *The Kashmir Times*, 64(242): 10.

Bahadur, J., 1999. Himalaya ecohydrology: An emerging topic. In: *The Himalayan Environment*, (Eds.) S.K. Dash and J. Bahadur. New Age International (P) Limited Publishers, New Delhi, Bangalore, Calcutta, Chennai, Guwahari, Hyderabad, Luckow, Mumbai.

Bhardwaj, B.L., 2005. Readers write: Melting of glaciers. *The Daily Excelsior*, 41(124): 6.

Bhran, R., 2006. Icy battlefield Siachen. *Springer*–The Kashmir Times Publications, January issue, pp. 1, 3.

Gupta, R.D., 2005. *Environmental Degradation of Jammu and Kashmir Himalayas and their Control.* Associated Publishing Company, Karol Bagh, New Delhi, India.

Khan, M.I., 2007. Changing climate and melting mountains. *The Kashmir Images*, 2(15): 5.

Morup, T., 2005. Seventy-five per cent biodegradable waste in Siachen including human waste. *The Kashmir Times*, 62(235): 1 and 4.

Shah, S.A.R., 2007. Glaciers and permanent snow–Kashmir losing fast its most precious natural resources. *The Kashmir Times*, 64(166): 11.

Sharma, B.K., 2000. *Environmental Chemistry.* Goel Publishing House, Meerut, Krishna Prakashan Media (P) Ltd., Meerut (UP), India.

Shiva, V., 2007. India: Globalisation and climate change. *The Daily Excelsior*, 43(162): 6.

Sidhu, G.S., 2006. Natural resource data for land use planning: Issues, constraints and strategies. *Indian Society of Soil Science Newsletter*. ISSN 091–8893, (21): 1 to 2.

Sinha, U., 2005. *Global warming* threatens human life. *The Daily Excelsior*, 41(276): 6.

Upadhyay, D.S., 1999. Meterological aspects of mountains, snow and ice. In: *The Himalayan Environment*, (Eds.) S.K. Dash and J. Bahadur. New Age International (P) Limited Publishers, New Delhi, Bangalore, Calcutta, Chennai, Guwahari, Hyderabad, Luckow, Mumbai.

Vasudeva, P.K., 2007. Gloom on the mankind: Climate change India's grave concern. *The Kashmir Times*, 64(186): 13.

Chapter 21

Landslides in Darjeeling District

V.C. Jha and P.T. Namgyal

ABSTRACT

Landslides are mass movements taking place over the earth surface on the downslope under the influence of gravity. Landslides have a considerable impact on the mountain ecology. Himalayan environment is under threat due to constant occurrence of landslides caused by natural and anthropogenic mechanisms. The present study deals with landslides and their impact on environment management in Darjeeling hills.

Introduction

Landslide is a very important type of mass movement which takes place over the earth surface on the downslope under the influence of gravity and the broken rock materials or rock wastes get accumulated at the base of the hills. The nature of the landslides depends upon the surface characteristics particularly relief, rock types, slope, vegetation and tectonic events which make equilibrium stage of the slope. This equilibrium stage is interrupted by various environmental factors particularly change in climate and loss of surface material but also the plant cover. It has become a serious geomorphological hazards which sometimes creates a very dangerous situation where the transformation of the rock materials takes place. There are various causes of landslides such as the nature of slope, the wetness of the surface, the structural control and the earth tremours.

Recently, in the domain of academic research the issues of environment and development are gaining much emphasis which was focused in the United Nations Conference on Environment and Development (UNCED) in Rio de Janairo in 1992. The dimension of environment varies from micro-environment to global environment and in nature it varies from physical environment to subjective environment. Generally, the geomorphological hazards are the result of both the natural and anthropogenic mechanisms. Landslides are mainly the result of exogenetic origin but their creation is from the various slope instability factors including the tectonic events. The nature's ecological balance is being increasingly intervened by various human activities directed towards achievement of control over the nature and to ensure material development in a true sense. The unjudicious human interference is affecting the environment adversly and is being observed in the form of different natural or environmental hazards.

Natural hazard is a condition of the environment that may exert damaging influence on human life, property and activities (Pal 1985, p. 55). Natural hazards are multidisciplinary in nature and can be typified on the basis of their origin. They are caused by exogenetic, endogenetic and anthropogenic forces (Nossins 1989, P. 183). Exogenetic type includes floods, droughts, landslides and avalanches; endogenetic type includes subsidence (from withdrawal of fluids, gases, peat drainage, hydrocompaction, mining etc.), reserviour quakes, collapse of structures (dams, dykes bridges, buildings etc.).

In this chapter an attempt has been made to find out the landslides affected areas in the Darjeeling district and their impact on environment management. This paper also suggests some measures for the checking of landslides.

Study Area

The Darjeeling district has been selected for the present investigation which is located on the northern most part of West Bengal in India. It is situated between 26 degree 31 minutes North-27 degree 13 minutes North and 87 degree 59 minutes East-88 degree 53 minutes-E. The district resembles an irregular triangle in shape. It is demarcated from Sikkim in the north by the rivers Teesta and Rangeet, from Nepal in the west by lofty Singhalila chain of mountains, from Bhutan in the north-east and in east by river Jaldhaka with Jalpaiguri district. In south the district is bounded by West Dinazpur and Purnia districts of West Bengal and Bihar respectively.

Physical Background of the Study Area

Relief

Darjeeling district as a whole is dominated by hilly areas. In the northern parts of the district, about one third part of the total area of Darjeeling comprising 3149 sq.km., is covered by the hilly areas. Southern part of the district is covered by plain lands. The highest absolute relief of the district is at Sandakpu hill (3638m) in the north western part of the district. The general slope of the study area is towards south of the district.

Geology

Darjeeling district consists of five major geological formations namely Gneiss, Dalings, Buxa, Gondwana and Tertiary. The outcrops of these geological formations form a series of bands running more or less parallel to the general relief trends of Himalayas and dipping on beneath the other into the hills. The most curious features of these formations is that the younger formations which always appear to underlie the older ones, thus the tertiary beds disappear under the Gondwana. The Gondwanas under the Buxa and Daling series and the latter under the Gneiss, the original order of superposition having been completely reversed by folding and faulting. The major formation available in the district is Daling series; other formations are available in horizontal bands running across the district from west to east.

Drainage

The valleys of the south-eastern side of Singhalila ridge are drained by the Mechi Balasan, and Mahananda rivers and all the remaining parts of Darjeeling district are drained by the Teesta and its tributaries, except the extreme eastern parts where the chief stream is Jaldhaka. Drainaage density in the study area varies from 2.65 km/km^2 (Balasan river) to 5.28 km/km^2 (Kali khola) and the drainage frequency varies from 3.73 $streams/km^2$ (Relli river) to 11.98 streams/ km^2 (Kali khola). Most of the rivers have high drainage frequency and high drainage density in the northern part of the district signifying rugged terrain.

Tectonics

The study area is influenced by several normal folding, faulting, jointing fracturing, cracks etc., taken place since precambrian time. Due to these tectonic events the original alignment of geological formation in the district is reversed to a great extent. The study area has been divided into five broad tectonic zones namely the Allochthon; which is subdivided into Upper, Middle and Lower Allochthon; Para Allochthon and Authochthon.

Climate

The study area is characterized by monsoonal climate. The average annual maximum temperature is 25°C and the average annual minimum temperature is 5.7°C. The rainfall starts from the month of June and continues till September. The average annual rainfall is noted about 110 inches. It is observed that the highest amount of annual rainfall is recorded at Kurseong (405 inch) and the lowest amount of annual rainfall is recorded at Kalimpong (225 inch).

Vegetation

Because of the heavy rain and moderate temperature, luxuriant vegetative cover is found in Darjeeling district. Broadly speaking, two major types of forests are found in Darjeeling district *viz.* The tropical forest upto an elevationof about 1800 m and the temperate forest with in an elevation of 1800 m to 3600 m. The main species found are *Sal, Dillenia pentagyan, Butea frondosa,* species of *Terminalia* and *Eugenia.* Along the river banks *Dalbergia sissoo* is fairly common accompanied by *Acacia catechu, Bombax malabaricum, Nandea cordifolia, Garuga pinnath* etc.

Landslides

Landslide is a form of mass movement in which rocks and debris moves rapidly down slope under the influence of gravity as a result of failure along a shear plane. The movement is often lubricated by water and tends to occur suddenly. It may result from undercutting of the slope earth tremours, heavy rainfall and the tectonic events which make the surface unstable A landslide may take a number of forms that includes slumping and rotational slips (Smith 1984, p.138). In the Darjeeling district the landslide is a very common geomorphological hazard.

Types of Landslides

In the study area it is observed that the most common types of landslides are Slumping, Rotational fall and Free fall. Besides these subsidence of land surface is also observed.

Slumping

It is a downward slipping of one or several units of rock debris usually with a back ward rotation with respect to the slope over which movement takes place (Thornbury 1984, p. 47). This type of landslides are not noticed in areas having moderate slope usually ranging between 30° to 40°. Frequently slumping takes place as several small independent unit resulting in that the surface of the slumped area has a number of step like "terracettes". Usually slumping is initiated due to under cuttings of the slope by human being for the construction of road ways. Usually on the moderate slope the thickness of soil is considerable. Due to undercutting the slope is accentuated and during rainy season the area having such characters experience minor landslides even though the amount of rainfall is not so high. The transported debris is usually deposited at the foot of the hills, however, frequently transported debris with certain interval of time gives structure of step like appearance though the structures last for very short duration. Examples of such type of landslides are noticed at Melli slide site and Rishi-Rangeli Padamchen Kupup road landslide site.

Rotational Fall

It is a rapid rolling or sliding of unconsolidated earth debris without backward rotation of the mass. This involves a turning movement that often leaves an upper surface on the failed mass inclined back into hill side. This type of landslide occurs with a concave upwards curved failure plane (rupture surface) passing through a thick and relatively homogenous bed of clay or shale (Hutchinson, 1968). They are more deep seated and the movement tends to be rotational on the slip surface. This type of landslide occurs in areas having steep slope and involves unconsolidated materials and becoming distorted with movement. Examples of this type of landslide site are noticed at Berrick slide site and Likhubir landslide sites.

In some cases a combination of slumping and rotational landslides are also noticed. In the upper areas normally rotational landslides are also noticed. In the upper areas normally rotational landslides occur and the rupture surface tends to be semicircular in shape, where water accumulates and maintains permanently high pore-water pressure, which triggers for slumping of the unconsolidated materials at the lower portion of the landslide site. This combination of slumping and rotational landslides are noticed at Berrick landslide areas.

Free Fall

This type od landslides occurs in areas where the slope is very steep and the materials fall from a vertical or over hanging cliff or bluff. This type of landslide observed in the study area is not exactly what we call rock fall. Here it involves the falling of fragmented rock materials usually in the form of soil under the influence of gravitation. Examples of such types of landslides are noticed in Ambotia landslide areas.

Subsidence of Land

It is another type of natural hazard observed in the study area. Subsidence is a downward settling of materials with little horizontal motion. The most common cause is the slow removal of materials beneath the subsiding mass. In the study area subsidence of land is observed at 17th mile, on road connecting Kalimpong and Algarah, Hermitage area in Darjeeling district.

Areas Affected by Major Landslides

The following areas of the Darjeeling district have been widely affected by landslides.

Berrick Slide

This land slide site is located 12 km on the Sevoke Teesta section of NH31A. This site is characterized by steep slope (around 50 degree). The principal rock types observed in this area includes orthoquartzite with band of dark grey phyllite. The quartz is dominant above the road but below the road, phyllite is more dominant. The Teesta river flows through just below the slide area causing toe erosin which has steepened the slope further. Phyllites at the lower areas disintegrate easily during heavy rainfall which results huge amount of debris flow downslope and steepening of slope. Quartzite in the upper side of the road is well bedded and highly jointed thus rendering quartzite freqmentation which falls under the influence of gravitation causing landslides delinking the routes from Siliguri to Gangtok and Kalimpong.

Likhubhir Slide

This slide area is located near 26km on the Sevoke Teesta section of HN31A. Geologically the area is characterized by grey schist and green slates with many joints. These rocks are soft and splintery. These rocks are soft and splintery. The area is characterised by uphill slope of 60 degree upto a height of 200 m and 70 degree to 75 degree of slope in the downhill slope. This slide is essentially a combination soil and rockfall caused by heavy rainfall.

Melli Slide

This slide has occurred in two areas 1) Slide I (near 33 km) and 2) Slide II (near 34 km) on NH31A. The area is characterised by folded greenish and grey phyllites dipping at 55 degree angle towards SE and NW. The slide is gradually advancing uphill and in the process has badly damaged, displayed and tilted catch water drain constructed in the past to arrest slide by channelising the surface runoff. Due to its location on the convex side of a bend in the river Teesta, the lower area of the slide area is affected by toe erosion which again has steepened the slope. In fact the slide seems to have caused due to rapid toe erosion, accentuated by rain and surface erosion.

Rishi-Rangeli Padamchen Kupup Road Land Slide

At several places along this road at 16 km and 17.3 km from Rishi Sewakhola (19.4 kms from Rishi) and Padamchen slide (2 km towards Keu Khola) massive landslide has occurred. These areas are characterised by sheared biotite gneiss, garniteferous muscovite gneiss, chorite muscovite schist, garniteferous gneiss with a slope ranging from 40 degree at Keu Khola slide to around 60 degree at slide at 17.3 km from Rishi. These rocks are characterised by many joints which causes splitting of rock materials. Whenever it rains, the fragmented rock materials ranging from soil to boulders get soaked with water and weaken the shear strength of the materials. Under the influence of gravitational force this debris slides downslope causing massive landslides.

Paglajhora Landslide

This site is located on the road connecting Darjeeling and Siliguri at Pagla Jhora. The area is composed of Golden-silvery mica schist, carbonaceous mica schist, garnetiferous mica schist, coarse grained gneiss. Many springs are found in the area which increase the moisture content in the soil coupled with heavy rainfall and the soil with mica particles laden with spring water, this area is invariably affected by huge landslide during every rainy season disrupting road transportation between Siliguri and Darjeeling for many days.

Rishi-Pedong-Algarah-Labha-Garubathan Slides

Along this section of road landslides have occurred at several places such as Rishi Algrarah section 57 km slide near Garubathaan, Rang Rang slide, Manual slide, Myang slide and Vong slide Here the uphill slope of the road is steep and rises for more than 60 m. The rocks are composed of garnetiferous sericite, chlorite schist interlayered with hard dark grey silicified schist and are characterised by joints, thin shear zones and drag folds which facilitates occurrence of landslide. Apart from these there are rocks which are overhanging above the road which endanger further rock falls during the rainy season.

At 17th mile road connecting Kalimpong and Algarah on the upper and lower parts of the road accumulated rock debris slides downslope during rainy season. Here the rocks are composed of Darjeeling gneiss and in parts schist is also present. These rocks are highly fractured. The area is characterised by presence of many small springs. Water seepage is also a characteristic feature of this area. Due to water seepage the underneath areas have been fragmented and lossened which has caused subsidence of land. All these factors are responsible for occurrence of landslides of this area.

It is observed that in all the above mentioned landslide areas along the Rishi-Algarah-Labha-Garubathan road the rocks are composed of mica schist, green and dark grey slates, biotite gneiss and are characterised by joints, foliations and steep slope varying from 40 degree to 60 degree. These factors influence the intensity of landslides. Apart from these seepage of rainwater along the hill slopes also induces occurrence of landslides in these areas.

Other sites of minor landslide affected areas are Lepcha-Jhora slide Rangpo slide Sewa-Khola slide, Padamchen slide etc. It may be generalized that most of the slides in Darjeeling hill areas have occurred at places where phyllite and schists are exposed. A few cases in granite gneiss and some in areas where quartzite are exposed. It is also noted that type of slides are related to the rock type.Areas having phyllite usually have slumping,schist usually cause slides and areas having gneiss and quartz usually experience rock falls. Most of the slides are restricted within the slope wash materials without invoking the bed rocks appreciably which is obviously thick and extensive in the phyllitic and schistose rocks.

Landslides and Environment Management

It is observed that in the study area the landslide is the most common and conspicuous environmental hazard prevailing almost occurrence of landslides every year. Occurrence of landslides not only disrupts the environmental equilibrium but also checks and retards the overall development of the study area. Therefore the management of such environmental hazards has become of prime importance in the pursuit of human development.

Environmental management comprises of a broad range of activities concerned with the human use of land, air plants and water (Cook and Doornakamp, 1990, p.1). In the broad sense environment means the rich natural diversity of the planet earth such as topography air, water, land, fauna etc. These myriads of factors are interrelated and are interactive to each other. The combined effect of the multitude of such interactions is to hold the environment in a state of dynamic equilibrium. Any change in any of the components of environment leads to disturbance of the equilibrium causing varied environmental hazards until a new state of dynamic equilibrium is achieved. This means if we let loose the environmental hazards to occur without checking it the situation gets aggravated and never could be able to maintain the required environmental equilibrium. Though we can not achieve a perfect and complete environmental equilibrium in this fast changing world atleast check and minimise those activities which trigger environmental hazards, for which conservation and management of environment is an urgent and important need.

Causes of Landslides

Landslides usually occur at unstable hill slopes, This factor stability and unstability is determined by the factor of safety FS, where

$$FS = \frac{\text{The sum of resisting slope failure}}{\text{The sum of disturbing factor}}$$

OR

$$FS = \frac{\text{Shear strength}}{\text{Shear stress}}$$

If FS is greater than 1.0 then the stability is likely to occur and is FS is less than 1.0 than the instability is likely to occur causing landslides.

Potrays a precise concept of factors involved in occurrence of landslides, processes involved and the preventive measures that can minimise the extent of landslide or check the occurrence of landslides.

Precisely landslides are caused due to unstability of land forms especially hill slopes. There are many factors which are responsible for unstability of hill slopes. The variation in degree of these factors effectiveness determines the intensity and spatial extent of slope failure. These factors are nature of relief, drainage, bedrock regolith, earthquakes, legacies from the past climate and human features (Cook and Doornakamp, 1990, p.121). Further Rudra, Bandhyopadhya and Bandhyopadhy (1982) while discussing landslides in southern sikkim have explained that the main causes of landslide are geological structure, rainfall distribution, intensity and duration stability of hill slopes and its relationship with cohesion and gravity toe erosion, relationship between the channel slope and the valley side, slope earthquake tremour, drainage system and road construction.

Since the physical and geological nature of the Darjeeling Himalayas, is identical to that of Sikkim Himalayas, it is observed that most of the causes of landslides in Darjeeling district are more or less the same as that of Sikkim Himalayas.

Geologically, Darjeeling district consists of coarse grain gneisses, Siwaliks, Gondwanas, Schist, Garnetiferous Mica, Schist, Carbonaceous Schist, Golden-silvery Mica Schist, Alluvium in the lower plain areas. The rocks are highly meta morphsed and most contain silicates which facilitate the sliding of debris downslope.

Apart from these, most of the areas in Darjeeling Himalayas are characterised by presence of fault, thrust, joints which makes the hill slopes more prone to slope failure.

The area receives heavy rainfall during the whole rainy season which sets in early June and lasts till September last. Darjeeling, Kalimpong and Kurseong receives annual rainfall amounting to 309.20 cm, 225.40 cm and 405.23 cm respectively. Due to heavy rainfall the dry weathered rock materials get lubricated and when water arrives rapidly within a soil body there is a tendency of high pore-water pressure to build up followed by slope failure.

The northern and the north-eastern parts of Darjeeling district are characterised by high mountains and hills. By virtue of its physical configuration the hill slope in these areas varies from average slope of 5 degree and above 25 degree. However, in places this varies from 50 degree to 60 degree. Generally along the high slope region soil formation is least but the bedrocks are characterized by fractures and joints which get fragmented during rain and increase the pore water pressure increasing the shear stress causing unstability of slope thus causing landslides. Because of the steepiness of hill slope any amount of loose material formed over these slopes, whenever gets soaked with heavy rain coupled with presence of silicates and mica particles in the loose materials (usually soil), slides downslope under the influence of gravitation causing huge landslides.

The drainages of Darjeeling district are characterized by rainfed and snowfed rivers which are actively involved in down cutting of river valleys and steepening of valley slope which increases the chances of causing landslides. Both drainage density and drainage frequency in Darjeeling district vary from moderate to high. High drainage density is a sign of such things as impervious strata, high rainfall, little vegetation cover and active incision, all of which may tend to increase the little vegetation cover and active incision, all of which may tend to increase the likelihood of mass movement (Cook and Doornakamp, 1990, P. 221). Seepage from a hillside especially along a spring line causes washout of fine sands and silts down the slope causing losses of slope stability. Besides toe erosion steepens the hillslope causing massive landslides afterwards.

Human interference into the nature induces the factors which lead to slope failure. In Darjeeling hilly areas remarkably felt effect of human intervention to the natural environment are construction works and deforestation.Many construction works involving excavation at the base of slope create steep slope and also remove the basal support. Unlimited deforestation along the hill slope causes direct impact of rain water with the bare surface due to which volume of moisture in the soil increases, binding of soil deteriorates, which finally results in loss of cohesion and hill slope instability causing massive landslides.

All the above mentioned factors are the basic factor for the occurrence of landslides in the district which has become a very common phenomena particularly during the rainy season.

Measures for Controlling Occurrence of Landslides

There are two complimentary approaches for controlling landslides namely general and particular approaches.

In general approach, the main emphasis is put on identification of landslide prone areas. To identify landslide prone areas the work is to be initiated from the spots of landslide occurrence as the reference point. This helps in understanding whether the landslide would be reactivated at that area and to understand the root causes of landslides in that area. Next step is to search for sites having similar properties but which for whatever reason have not experienced landslides. Many techniques have been developed with the passage of time to accomplish the above mentioned exercise of identifying landslide areas. Among them, proforma check list is most important which involves collection of various data on the factors controlling landslides. Proforma check list actually is a list of various factors causing landslides such as relief, drainage, bedrock, regolith, earthquakes, legacies from the past, climate, human features etc. Depending upon the values of these factors areas having higher or lower tendency of experiencing landslide is determined. Other techniques are data collection on a grid basis, landslide inventory mapping, geomorphological mapping etc.

Particular approaches for controlling occurrence of landslides includes construction works, plantation of plants species having strong soil binding capacity, checking of human interference within a range of fields having no geomorphological risks. The engineering construction works involves construction of breast wall, barbed wall, wall cuttings, spurs, toe protection walls, water catch to channelise runoff, water drains and chutes, heavy sewage wall, check walls, etc. For instance, in the case of Berrick slide site following particular remedial measures should be taken for checking landslides in that area.

1. Removal of loose rock materials from the slope face and from the cracked rock ledges near the crown of the slide.
2. Provision of two spurs at the right bank of Teesta river to protect the slope from toe erosion.
3. Construction of atleast two berms up the road with deep side drains leading to two chutes across the slide area.
4. Protection of phyllite rocks especially down the road by shortcreting.

In case of Likhubhir landslide site, protective measures such as breast walls, barbed wall, wall cutting have been adopted. To check landslides at Pagla Jhora site, sausage wall and breast walls have been constructed. But it is obersved that the construction of these engineering structures has not always beeen successful in arresting landslides.

It is observed that along with the construction of protective physical structures, a thorough survey of the areas affected by landslides and identification of areas having similar characteristics as that of landside affected areas should be conducted. For a systematic analysis of the area, drainage basins should be considered as unit area for study and survey. Appropriate techniques to suit the deficiency of each and every area should be adopted. Restriction of deforestation and revamping of afforestation should be encouraged with plant species having strong soil binding capacity such as Babio (*Enclopsiobinita*), Amliso (*Thysonaleaner agrostis*), Gueatmala (*Tripsicum laksum*), Napier (*Pennesituim perpurenum*), Bankodo (*Dechylocterium aegiticum*) should be planted. These above mentioned measures will certainly be meaningful for restoring the ecological balance in the study area.

References

Chorley, R.C., *et al.*, 1984. *Geomorphology*. Methuen and Co. Ltd., London.

Cook, R.U. and J.C. Doornakamp, 1990. *Geomorphology in Environment Management*. Oxford University, Press, New York.

Leopard, L.B. *et al.*, 1964. *Fluvial Process in Geomorphology*. Eurasia Pub. House Pvt. Ltd., New Delhi.

Nossin, J.J., 1989. Aerospace survey of natural hazards the new possibilities. *ITC Journal*, p. 183–188.

Namgyal, P.T., 1991. Darjeeling district : A study in geomorphology and environmental management. *Unpublished Ph.D. Thesis*, Visva Bharati, Santiniketan.

Thornbury, W.D., 1954. *Principles of Geomorphology*. John Wiley and Sons Inc., New York.

Index